2011

The Best American Science Writing

THE BEST AMERICAN SCIENCE WRITING

EDITORS

2000: James Gleick
2001: Timothy Ferris
2002: Matt Ridley
2003: Oliver Sacks
2004: Dava Sobel
2005: Alan Lightman
2006: Atul Gawande
2007: Gina Kolata
2008: Sylvia Nasar
2009: Natalie Angier
2010: Jerome Groopman

The Best American

SCIENCE WRITING

EDITORS: REBECCA SKLOOT AND FLOYD SKLOOT

Series Editor: Jesse Cohen

An Imprint of HarperCollinsPublishers

Contents

Introduction by Rebecca Skloot and Floyd Skloot

"You can observe a lot just by watching."

—Yogi Berra

SCIENCE WRITING HASN'T ALWAYS been the Skloot family business. Until it happened, there was never a moment when it seemed likely that either one of us—let alone both of us—would devote our lives to writing about science.

Skloot the Elder (a.k.a. Floyd) grew up wanting to be a professional ballplayer, despite his lack of size and talent. Most of his dealings with science involved broken bones and head injuries. As an English major in college, the only Ds he earned were in geology and economics. After more than forty years as a poet and fiction writer, he eventually came to science—or more accurately, science came to him—when a virus targeted his brain and left him disabled, his neurological and cognitive functions seriously impaired. He started writing about science out of an intense need to understand what was happening to him: he began explaining it to readers as a way of understanding it himself.

Skloot the Younger (a.k.a. Rebecca) did the opposite. One of her earliest toys was a Fisher-Price medical kit she used to identify and treat an endangered population of stuffed animals. She was determined to be a veterinarian (even after failing multiple science classes because she didn't show up), so she went to night school, earned a biology degree, and worked in pathology and neurology labs, veterinary morgues, emergency rooms. Literature and the arts, despite— or maybe because of—her father's involvement in it, seemed irrelevant to what she wanted to do: science. It wasn't until she was in college that, to fulfill the school's foreign-language requirement (!!), she took her first creative writing class. There, in response to an assignment to "tell the story of a place," she wrote about the freezer in the veterinary morgue, about the ethical questions raised by the bodies filling it daily. As she listened to the intense debate that story inspired in her fellow students, she fell in love with writing's ability to expose people to areas of science they didn't know existed, to get them thinking and talking about science.

People often talk about the arts and sciences as separate worlds inhabited by vastly different species: right-brain people in one, left-brain in another. But in our experiences, the arts and sciences are more alike than not: both involve following hunches, lingering questions, and passions; perfecting the art of productive daydreaming without getting lost in it; being flexible enough to follow the research wherever it leads you, but focused enough to never lose sight of your larger direction and goals. There's an alchemy that occurs when art and science come together, when the tools of narrative, voice, imagery, setting, dialog, are brought to bear on biology, chemistry, physics, astronomy, mathematics, and their various combinations. And though we came to science writing from opposite directions, we arrived with convergent ideas of what makes science writing important and effective. This seems fitting, since for us, good science writing—as with good science—is dependent on the uniqueness and creativity of the approach.

Science writing isn't always easy to define, because it comes from many different fields, angles, and experiences. Sometimes the science is overt, other times subtle, an essential backdrop to a larger human story. So we approached this collection with a wide definition of "science writing," one that accommodated our shared belief that science is all around us, not just in labs and schools and hospitals. It exists in living rooms, kitchens, grocery stores, and meeting rooms; on crowded streets, ball fields, and plates of sushi; it's in Dumpster dives, walks in the neighborhood, patterns of events, and casual statements overheard. Science is everywhere, and we wanted stories that showed just how present and personal it is, how intimately it affects everyone's lives.

So we took a broad view of the science part of science writing, with its everyday subjects like Michael Rosenwald's harrowing and hilarious confessions of his problems with hoarding ("The Mess He Made") along with stories of traditional research and discovery, like Julia Whitty's exploration of how and why the most threatening and enduring damage from the BP oil spill will be occurring at the deep layers of the sea ("Deep Secrets").

We wanted to showcase the wide range of subject, voice, approach, and style represented in good science writing, and we were drawn to work that pushed against the conventions of form or topic. So our selections include a *profile* of the evangelical Christian geneticist Francis Collins and a medical *memoir,* a work of *long-form journalism* about cybersecurity, and an *immersion piece* from the edges of science fiction where neurology, computer science, and philosophy of the mind come together with the wildest imaginings about human function. There is a *historical investigation* of how psychotherapists used LSD to treat Hollywood celebrities in the 1950s, a *future projection* of the longest possible home run, and an *opinion piece* about the challenges and responsibilities that come with writing about science.

There's also an *exposé* on why so many TV weathermen don't believe in global warming and the dangers of the media speaking recklessly about science; plus a *report* of what happens when physicists turn their attention from subatomic particles to look at "important information hidden" in the numbers of terrorist attack victims.

While we found great stories in outlets that reliably publish fine science writing—*The New Yorker, New York Times, OnEarth, Discover*—we also found them in publications not best known for science writing: *Columbia Journalism Review, The Week, Playboy, Vanity Fair*. We also found good science writing on blogs, including many written by scientists themselves.

Historically, there's been an enormous divide between scientists and the general public, but today, more and more scientists are reaching across that divide by writing about their work and interacting with the public on blogs. We're excited to see this, because we believe when done well, science blogging can play an essential role in increasing understanding and communication between scientists and the public.

So what makes good science writing? Humans have been learning through stories since our beginnings, whether from cave paintings or oral history. So we believe the best science writing presents information clearly and accessibly while also telling stories that show readers how science impacts them, why it's essential to life and culture, why they should care, and why they should learn about it. When we talk about storytelling, we mean writing that uses literary techniques: scenes, narrative, voice, characters, dialogue, setting, and action.

Science happens to people, and is done by people, so we wanted the stories we chose to reflect the powerful presence of character. We wanted those who were being written about—the neuroscientists and geologists and physicists, the explorers, professors, "fermentation fetishists," media figures, isolated sufferers of disease, the doctors and deep-sea divers, subjects of research, even the robots or the carp—to be brought vividly alive by the writing. We wanted to see how characters carried themselves, hear how they spoke and understand how they thought, and we didn't want to be able to forget them.

Whether the piece focused on a walk in the desert, or the "Conficker" computer worm; whether it was about medical research, compulsive behaviors, strange appetites, or the possible roots of human cruelty to animals, there had to be intense, passionate engagement on the part of the writer, a sense that the words matter as much as the information conveyed. We wanted surprising insights that made us reread to be sure we had it right ("the future of climate change policy in the United States rests to a not-insubstantial degree on the well-tailored shoulders of the local weatherman"). We wanted turns of phrase that startled us (the "greasy waves" in Barataria Bay after the oil spill, or the child weakened by muscular dystrophy who couldn't stand, but would "blob down"), structures or language or viewpoints unique to the material being presented rather than generic or academic. We found pieces that made us laugh, made us angry, made us cry (well, made Floyd cry, which admittedly isn't very hard to do).

We also wanted obsession—something that drives scientists to important discoveries, and writers to important stories, committed stories that keep going deeper into their subjects. Nearly all the pieces in this collection combine everyday subjects and traditional research in a way that charges the writing with intensity because the author is deeply involved in the story, sometimes very personally. Cynthia Gorney's in-depth study of the conflicting research on hormone replacement therapy, "The Estrogen Dilemma," is fired by its personal relevance for Gorney, as is Katy Butler's reporting on pacemakers extending the life of late-stage dementia patients in "What Broke My Father's Heart."

Science writing is personal not just for the writer and subject, but also for the reader, so there were certain topics we gravitated toward because of our own passions and experiences. We're big into the science of food. We've been cooking together since Rebecca was two and demanded an explanation for why the oatmeal bread dough she was kneading was so much stickier than wheat dough. For obvious reasons, we're also drawn to stories about animals, sports, and the brain, both how it works and how it doesn't. Also pieces about human relationships with computers, since Floyd is so stumped by the things that he constantly calls Rebecca for help with them.

But we also found ourselves drawn to stories from everyday life that had no obvious personal significance for us—until we read the stories. Now we're very attuned to the worldwide plague of coal fires burning in abandoned mines, and to the power of paying attention to data and its trends (Dad, you empty your computer's trash *how* many times a day? Becka, you drink *how much* coffee?).

We ended up with many stories about medicine, and the human mind. Also many about health: of humans, of the planet and other species, of our communities—especially the way we communicate within and among them—and even the health of the field of science writing.

In "The Trouble with Scientists," Deborah Blum points out— with charming understatement—that "the culture of the 'real' scientist who exists somehow separate from the rest of us has not been a boon for public understanding or appreciation of science." It's no longer news that American education in math and sciences is in decline, and scientific illiteracy is rampant. This failure to educate needs to be addressed and, as Blum says, science writers are in the best position to help us "move forward in improving public understanding of science." They bear a responsibility for communicating facts accurately and clearly, for avoiding the sensational, for educating readers and a public increasingly unversed in (and sometimes even afraid of) science. They are, as Blum says, the "public communicators of science."

It's an honor to be the first collaborative editors in the twelve-year history of *The Best American Science Writing* series. Though we've wanted to for many years, we haven't—until now—published together. It feels appropriate that our first collaborative publication should be a gathering of the year's best science writing, since science is ultimately a deeply collaborative field, its practitioners sharing information and findings, replicating and legitimizing one another's discoveries, challenging one another, and in the best situations inspiring one another. Science writers do this too. And certainly the work included in this year's anthology reflects this sense of cooperative endeavor, writers following scientists around, going in the field to observe and report, bringing diverse individuals together in a conversation taking place on the page.

The Best American Science Writing 2011

MICHAEL S. ROSENWALD

The Mess He Made

FROM THE *WASHINGTON POST*

The writer Michael S. Rosenwald thought he was a slob. The fact that he had piles of "stuff" throughout his home, or old mail, or bags of untossed garbage, suggested something more clinical: that he was a hoarder.

M Y PARENTS RECALL THAT MY TEENAGE ROOM WAS such a disaster, the piles of clothes and old newspapers so high, that our dog Ozzie considered it equivalent to the back yard and used it accordingly. Ozzie was clever enough to open closed doors, so my parents installed a chain lock on the outside. The chain naturally prompted questions from visitors, the most tactful being: "Why are you locking your son away?"

Nearly 20 years later, my high school girlfriend cannot shake the memory of being surrounded by my piles. "I remember your room smelling so bad I would seriously breathe out of my mouth until I could somewhat get used to it," she told me when I tracked her down on Facebook. "I can't describe it, but I'm sure if I did one of those tests with a blindfold, I could pick it out even to this day. Maybe a mixture of old sneakers and dirty clothes and rotting food all mixed together."

My parents once moved all of my stuff to the front lawn, hoping the embarrassment would reform me. Ingenious. Didn't work.

My problems accelerated after I left for college. My freshman roommate, who eventually became one of my closest friends, only recently told me that he had requested, but was denied, a roommate change after he was unable to safely walk from one end of our room to the other without slipping on a pile of newspapers, magazines, books or unopened mail.

Living alone made matters worse. When I was in graduate school, burglars stole a laptop from my apartment. The detectives, two women who reminded me of Cagney and Lacey, took only a few seconds to offer their first investigative finding: "Wow, your place really got ransacked." I explained that nothing in the apartment had been touched, including stacks of several years' worth of newspapers, and that I hadn't cleaned up because I had wanted to preserve fingerprint evidence.

There was silence. Then one of the detectives said, "We're calling your mother."

I said, "She knows."

You may be surprised to learn that I am married. I should confess that my wife, Megan, was not briefed on any of these tales when we first met in Boston, and I made sure she didn't learn of my special qualities until I had charmed her extensively. "I felt you kind of deceived me when we first met," she told me recently. "You had your car professionally cleaned, a friend picked out your clothes, and you even hired a maid to clean your apartment before I came over the first time."

For reasons I still can't totally explain, she not only agreed to move with me to Maryland a few years later but also said yes, without a twitch, when I proposed.

Still, she wondered. "I remember going through your books before we moved," she said, "and finding two or three copies of the same book. Who does that?"

Megan's question led me to confront myself: Am I, as she puts it, the laziest, most nauseating slob in modern U.S. history? Or is something else going on—something more complicated?

Am I a hoarder?

LIKE MOST PEOPLE, WHEN I think of hoarding, the images that come to mind are the horrific scenes of uninhabitable homes that enter our living rooms during Sweeps Week. We watch these TV tales in the same way that we slow down for multiple-vehicle pileups on the Beltway. A couple of years ago, in an episode titled "Inside the Secret Lives of Hoarders," Oprah Winfrey visited a Rockville couple whose 3,000-square-foot home was overflowing with 75 tons of garbage. I went looking for the clips the other day on Oprah's Web site, and the page about the show shouted, "Uncover what's behind a hoarder's closed doors!" I felt my stomach turn. The exclamation point, to me, screamed: "Freak show here. Step right up." For many Americans, these are hoarders, no further details needed. But as I now know, that's not the whole story.

After soliciting recollections of my slobbiness from friends and family, I looked into the scholarship on disorganization and hoarding. The first book I came across was co-authored by Randy Frost, the world's foremost hoarding expert. Titled *Buried in Treasures: Help for Compulsive Acquiring, Saving, and Hoarding,* the book prompted a double take from the cashier when I paid for it at Barnes & Noble. Usually, I don't ask for a bag. This time, I did.

The book includes a questionnaire Frost devised to identify

hoarders. Reclining in my living room La-Z-Boy, I pulled out one of my favorite fountain pens and took the test. (I have hundreds of fancy pens, vastly more than I could ever use.) One question was: "How much does clutter in your home interfere with your social, work or everyday functioning? Think about things you don't do because of clutter." Our dining room table and its chairs are totally covered with my piles of papers and at least a dozen bottles of fountain pen ink, so the idea of having people over for dinner or even to watch a football game is rather exotic.

Another question: "To what extent do you have difficulty throwing things away?" Answer: I tell my wife I am throwing things away, but, really, I just hide stuff in other places.

"How strong is your urge to save something you know you may never use?" Before leaving Boston in 2004, I found a box of unopened mail—catalogues, flyers, bills and letters from collection agencies—dating to 1993.

I totaled my answers to the test's 15 questions. My clutter score qualified as "severe." My "difficulty discarding" score qualified as "severe." My "acquiring" score qualified as "severe." I looked up in the direction of our dining room table and thought, Uh-oh.

FROST HAS A CORNER OFFICE in the humanities building on the campus of Smith College, a women's school in Northampton, Mass. He invited me to visit so I could be psychologically dissected in his advanced seminar on hoarding. I was to be the guest specimen. When I arrived, he was straightening his desk, which was already tidy. Frost is 6-foot-5, built like a basketball forward, with a tightly groomed mustache and much less than a full head of hair. He is charming and soft-spoken, two qualities that probably ingratiate him well to the thousands of hoarders whose homes he has visited for his studies.

Frost came to Smith in 1977 to teach abnormal psychology, but it wasn't until 1991 that he stumbled on hoarding. A student was discussing ideas for her term paper. She wondered about people who could not throw things away. Could she write her paper on this topic? Frost did not think the behavior, while abnormal, was widespread enough to justify a term paper. "Hoarding is something you don't see very often, and there is no literature on it," he said to her. But the student persisted. She mentioned the Collyer brothers, the infamous pair who died under 130 tons of junk in their tiny Manhattan apartment—they are the subject of a new novel by E. L. Doctorow—and how her mother often told her to "clean up your room so you don't end up like the Collyer brothers."

Frost told his student to place an ad in the newspaper to see if any hoarders might help her with firsthand accounts. He expected a few calls. They got 100. "It's been like a runaway train since then," he said. He recently co-authored a new book on the subject, *Stuff: Compulsive Hoarding and the Meaning of Things*. Frost said current studies show that from 2.5 percent to 5 percent of American adults suffer from some form of hoarding. In 1991, Frost and his student knew only of those 100 callers, and they surveyed them in the first major study of hoarding.

In many homes, they found, stuff was piling up—not necessarily overtaking homes but rendering many important functions impossible. One subject's house had "a series of maze-like paths through rooms piled to the ceiling with miscellaneous objects." Another had no clutter visible, but the basement and attic had hundreds of boxes stacked in rows up to the ceiling. Kitchen tables became storage pallets. Living rooms became labyrinths. Spare bedrooms became flea markets. But why? Frost and his student's quizzing of these initial subjects turned up some tantalizing clues.

Saving allows the hoarder to avoid making what they view as risky decisions. For the past decade, I have bought, nearly every other year, the same exact pair of brown and black Timberland loafers. When I'm done with them, I never throw them away. I probably have six or seven pairs stashed in various places. Something about throwing away shoes makes me uncomfortable—the same way I feel when looking at a three-year-old *New Yorker* magazine that I stuff under the bed or when I move a three-month-old Best Buy catalogue from the dining room to the basement.

These behaviors typically emerge in adolescence. My mom remembers that my years spent waiting tables were profitable for her because she could always count on finding dollar bills around my room. Lift up a stack of papers, find five bucks. A lottery of sorts. She barred the cleaner who came to our house once a month from entering my room—not because she wanted to keep the money for herself, but for fear that if the woman started cleaning in there, she'd never come out. My room often provoked arguments between my parents. My dad would say, "How can you let him live like this?" My mom would say: "His room. His choice."

Frost's research also showed that many hoarders have close relatives who behaved similarly, suggesting a genetic component to the phenomenon. That would be my father's father, Sam, after whom we named my son. Sam, a widower, lived alone. He ate out for every meal. Nobody ever visited his apartment. When we picked him up, we drove up, honked, and, a few minutes later, he would slip out his front door.

After my grandfather died, my father entered his apartment and was astonished. It looked very much like my house does now. There were, for instance, hundreds of old *Reader's Digest*s scattered around. "They were everywhere," Dad told me.

I told Frost I was the same way: "I know I'm never gonna use all this stuff I save, but I still keep it."

He leaned forward and said, "We hear that over and over again."

I told Frost I could sense my wife becoming increasingly frustrated with my piles. He said marriages with hoarders often fracture because the collectors cannot tolerate the boundaries their spouses set. "It sounds like that's the most dangerous thing for you right now," Frost said. "If you become more rigid about this or if it becomes too much for her, then it's gonna be worse."

I told Frost about my son. He is only 2, but this behavior pattern needs to stop—somehow, some way—so he doesn't follow my path, and his namesake's, too. I explained that I feel desperate to give my boy whatever nurture he needs to head off what nature might bring his way. Now that Sam is tall enough to see the top of the dining room table, I wonder: When he looks at it, what does he think?

OVER THE YEARS, THERE HAVE been interventions. Friends helped me gut my bachelor apartment, hauling out enough garbage bags to move the Hefty company's stock higher on Wall Street. One set of friends chose to wear gloves. An old girlfriend once cleaned out my apartment while I was unconscious in bed after wisdom tooth surgery. This caused me more stress than the throbbing in my mouth.

God, via clergy, has also intervened. Before Megan and I got married, our rabbi held three counseling sessions with us, two of which we spent talking about Megan's contempt for my living habits. Megan told the rabbi that my slobbiness made her worry about our future. I told him what I'd always told my parents and others who have confronted me: The stuff is mine; it doesn't bother me; it's on my side of the room; just ignore it. The rabbi didn't so much try to offer solutions as to air out the issue. He asked Megan whom she blamed. I was surprised when she replied, "His mom." She recently told my mother the same thing, and though laughter followed, it was of the nervous sort.

I raised the issue separately with my mom (initially via e-mail, because I lack guts), and her reply was: "That's just great. Now I look like a bad mother." I felt like a bad son. Seeking to defend her to herself, I pointed out in another e-mail that if she was responsible, then why wasn't my sister, who is a total neat freak, just like me? We are 16 months apart and were raised in the same house, at the same time, under the same regime.

But (and this is a big but) if my mom is not to blame, that would implicate my genes in some way. And if my genes are responsible, again, why is my sister neat? The roots of my problem are more complicated than a simple designation of nature or nurture, something Frost and other researchers have yet to pin down.

No matter who is to blame—if anyone—Megan has now suffered through years of dealing with my piles, and though she isn't pleased, to say the least, about this story, she was thrilled when I started to look into other forms of intervention, particularly since her threats of "clean this up before I get rid of all of it" have not changed me.

They have, however, kept me from tilting back into total disarray. Without my wife, the piles would grow until they took over the house. There would be at least four years' worth of newspapers in the kitchen—on the counters, the floor, the table, under the sinks. We live an endless loop: She complains; we argue; I clean a little; piles grow back. Repeat. Frost, in *Buried in Treasures*, suggests calling the pros if you feel overwhelmed by the issue and your friends or family can't help you get things in order. Another reason to seek help: if anxiety or depression is getting in the way.

A few ideas for de-cluttering have surfaced over the years. A storage company offered to send me a large container to move all my stuff into. The company's pitch: "It can be your new man cave. We at Units Mobile Storage will bring a unit right to your driveway. . . . Set up a TV and thoroughly enjoy life surrounded with all your stuff without your wife having to live and breathe it every day. It will cut down on the nagging and may indeed save your marriage!"

Megan nixed the idea. "First of all, it's ridiculous," she said. "Second of all, our homeowners association would throw us out."

Another idea came from Bernie Kastner, an Israeli psychologist and handwriting analyst who, upon hearing my story, offered to study my handwriting to find ways to help. I sent him a one-page handwriting sample. A few days later, he sent me a three-page analysis of my personality that was so accurate as to be frightening. The report: "When directly confronted and threatened with the possible consequences of his actions, he may dig in his heels and become even more insistent on doing what he wants."

Me: Are you taking notes, Megan?

Kastner and I chatted on the phone. He suggested turning my messes into a game. Bet money that I won't keep clean. If it's a game and I stand to win some cash, that should clear up the problem. Brilliant, I thought. We hung up, and then I remembered that Megan had once tried something similar to get me to make the bed. She would grade me on my bed-making efforts. If I scored high enough each month, she would treat me to a steak dinner. There is literally nothing I won't do for steak—except, it turns out, make the bed.

Then Mo showed up. Mo Osborn is a nurse turned professional organizer. And not just any organizer but a member of the National Study Group on Chronic Disorganization, a group of 200 organizers who help people like me. Mo is petite and bubbly and utterly charming.

I led her on a tour of the house. First stop, the dining room table. I admitted: "I'm not going to lie; this stuff that I have on the chairs over there, that was stuff on the table that I didn't have room for anymore, so I just moved it to the chairs." Mo's advice: Get a small bookshelf to keep nearby. I get to have my stacks, but they would be out of the way.

Up to the bedroom. Mo looked at the leaning tower of books and magazines. I said if I roll over too violently while sleeping—I am the violent rollover type—Magazine Mountain crumbles. This has happened a few times, generally between 3 and 5 a.m. Mo's idea: Donate the duplicate books and the ones I won't ever read to a prison book program or the State Department's reading abroad program. The theory: If I know they are going to a good purpose, that will help break my attachment. Also: Put a recycling bin next to the bed. If I sort right away, instead of waiting for the piles to grow, I will under-cut my tendency to save.

Mo was full of great ideas. I was excited. I told Mo that my wife would call her to say thanks, that she had given me a path, both liter-ally and figuratively, to a cleaner, less cluttered life. It would be like living in a hotel suite. "I'll report back to you soon," I said.

But the tools have to be used for them to work, and only I could employ them. A week went by. Mo called to see how I was doing. I didn't call her back. I sent her a note saying that the *Washington Post* was dispatching a photographer to take pictures of my messes, so I couldn't touch anything. That was essentially true, though I could have bought the bookshelf and tidied up a smidge. The photogra-pher came, and a few more weeks went by. Of the many excellent suggestions Mo offered, I had implemented exactly none.

WE SAT IN A CIRCLE in a small classroom down the hall from Frost's office. I was tense. My audience was students specializing in the study of abnormal behavior, and I was the abnormal one. As I introduced myself, I stumbled over my words. But as I talked more, offering details about my slobbiness, I grew more comfortable. I felt like I was unloading a secret, a burden. The dozen students of Psy-chology 354, Seminar in Advanced Abnormal Psychology: The Meaning of Possessions, were there to help me, not judge me. In that setting, I began to sense, for the first time, why so many interven-tions had failed.

"My wife and I were in a bookstore recently," I told the students, "and she said, 'I don't know why you're shopping in a bookstore; you have accumulated a bookstore next to the bed.'" A few students gig-gled. That pleased me, probably, I realized, because my identity has become tied up with being a slob, just as Woody Allen's is tied up in being a hypochondriac. The students were shrinking me, as the saying goes, but I was also shrinking myself.

"I had garbage bags everywhere," I said, detailing my attempts to clean my apartment before I left Boston. "One of the garbage bags happened to have a light bulb in it for some reason, and I stepped on it with my bare feet and needed surgery." I waited for the students to laugh. They didn't. One gasped. Maybe this wasn't something to laugh at. Maybe, all along, there has been an audience of one: me.

Later, I would learn from Frost that I keep my stuff on tables and in piles because having everything in plain sight provides comfort and, in a sense, a form of organized disorganization. If I can see it, I know it's there. That was the practical explanation. But as the students questioned me—about the pleasure I feel acquiring stuff, the anxiety I feel tossing it—I sensed that there was something deeper, more philosophical. And it was this: All of the stuff I pile up is a sort of second body, my twin. I am Michael Rosenwald, and those piles— the books, magazines, fountain pens, inks, newspapers, everything—are also me. The more I have of it, the more I am me. Up there in front of the class, I was beginning to confuse myself, and then I felt as if I might cry.

I blurted this out to the class: "What would I be without it all?"

Frost said: "What am I without my things? That gets to this whole issue. A sense of identity. What am I without my stuff? What's happened over the years is the stuff has somehow invaded your sense of self, your identity, because without it you feel like you don't know who you are."

That clicked. I will buy books more than once because I can't remember if I bought them or not, and I feel like if I don't get them, I will never have a chance to have them again, and I need those books. I need them, it turns out, to keep up with my concept of myself. I recently bought a copy of a magazine that I had bought two weeks before. When I got home, there it was, the same magazine, on my nightstand. How could I have done that? What a waste of money. But I did it. Had to have it—again.

As the class stared at me, probably wondering what planet I had arrived from, my sense of ease slipped away. I wanted to be anywhere but in that classroom. I crossed my legs, then uncrossed them. I took a drink from a bottle of water. Then it hit me: What if I really am a hoarder? I shot a quick glance at my cellphone to see how much time was left in class—45 minutes. Yet it was also a moment of deep clarity.

Then a student asked the question I had secretly been hoping for: Have you told your wife that you think you might be a hoarder? The money question! My chance at innocence! I have a condition, Megan. I'm not a slob. I'm a hoarder.

I sat straight up, cleared my throat and delivered my response, which I had been rehearsing in my head for days: "I think it is hoarding. She thinks I'm lazy. So there's a huge disconnect. She's also a physician; I didn't mention that. She's a family doctor, so she sees a lot of mental health issues. Her perception of hoarding is the Oprah image, which is, let's go into somebody's house and see the things toppling over them. What I've learned is that, yes, that is hoarding, but there is another way of getting toppled over on yourself and your relationships."

The rest of the class felt like a blur. I was there but not there. I was in my bedroom, on top of that pile, looking back at myself lying on my bed, staring at it all. I was on the dining room table, looking out at myself from under a pile of newspapers. I was in a bookstore, watching myself walk around, looking at books that made me feel more like me.

"We have run out of time," Frost finally said. "Thank you, Mike. This is very brave, very courageous."

Then he whispered in my ear: "Let's go back to my office. I want to make sure you are okay." I told him I was. That night, I barely slept.

I TOOK MY SHIRT OFF. My wife asked, "What are you doing?" I said I was getting ready to clean.

"Does your shirt have to be off to clean?" she asked. "I'm thinking I might sweat," I explained.

She said, "I hope you do."

I took three Advils. I assumed a headache was inevitable. We stood in the dining room, clearing what remained of the junk I had begun disposing a couple days before.

I said, "I need to set my lineup for my fantasy football team."

She answered, rather loudly, "You're doing what that book says I shouldn't let you do."

Megan was referring to Frost's book—the self-help part, where he helps reformed hoarders overcome what he calls "the bad guys" that get in the way of organization. This was the "It's not my priority" bad guy, found on page 139: "If you find that other things start to seem more important to you than sorting and organizing, stop and reassess your goals and priorities. Are these other things true emergencies that you really must attend to right now?" I told my wife the fantasy team lineup was important. She reminded me I was in last place in my league.

We moved upstairs to the leaning tower of books and the nearby piles of magazines. We made two new piles: go-away and keep. We used the bed as our sorting station.

I asked, "Do you want to turn on the TV?" She said, very lovingly, "I want a new husband."

We were off to a splendid start. She began digging into the pile next to my bed and said, "Oh, a pair of shoes."

I said, "I was looking for those."

She said nothing.

Then we started through the books. Of the first five books we examined, three were identical to volumes I already had downstairs. The very thought of putting one of them on the go-away pile gave me heartburn. Another Frost bad guy had arrived—the "Unhelpful beliefs about your stuff" one. "I feel so attached to these things!" states the definition of this bad guy. "But all this stuff is useful!" I turned to page 149: "How many do I already have, and is that enough?"

I said to myself: "Two." I also told myself that the book I already had was the same, word for word, as this one. It went in the go-away pile.

This didn't feel as bad as I had thought it would. I kept telling myself, This stuff isn't me. If it all disappeared in a fire, my body would not implode, my identity wouldn't turn to ashes. I would emerge, walking out the front door with soot on my face, the same person I was before the flames, only without the stuff. The stuff was not me, the stuff was not me—it felt like some self-help mantra. The more I told myself that story, the easier the tossing became. We went on like this for an hour in the bedroom. For every book I kept, I let five more go. Every time I showed signs of indecisiveness, my wife said: "Do you really need this? Are you going to die without it?" The result was three boxes for the Salvation Army.

As I carried the boxes to the car, I thought about a question I was asked in Frost's class: "What's your fantasy about how you want your living space to look?"

I said: "I love hotels, and when I go into a hotel room, I love how clean it is, and I love the orderliness of it. I guess most of all I just don't want to be nagged anymore. I don't want to be stressed out by it anymore."

When I went back upstairs, my nightstand was clean, and the floor around my bed revealed carpet I hadn't seen in months. It didn't look like a hotel room, but it was close, at least to my eyes. We cleaned up my stuff throughout the house. It took all day and into the next one. I told my wife how much I liked everything clean, and she reminded me that I have cleaned before, only to relapse. I vowed this time would be different.

She said, "I hope so."

I said to myself: "I know why I do this now. I've got this figured out."

Two weeks later, the piles were back.

KATY BUTLER

What Broke My Father's Heart

FROM *THE NEW YORK TIMES MAGAZINE*

> *A pacemaker can save a life. In the case of Katy Butler's father,
> however, it only prolonged his misery—and that of his family.*

ONE OCTOBER AFTERNOON THREE YEARS AGO WHILE
I was visiting my parents, my mother made a request I
dreaded and longed to fulfill. She had just poured me a cup
of Earl Grey from her Japanese iron teapot, shaped like a little pump-
kin; outside, two cardinals splashed in the birdbath in the weak
Connecticut sunlight. Her white hair was gathered at the nape of her
neck, and her voice was low. "Please help me get Jeff's pacemaker
turned off," she said, using my father's first name. I nodded, and my
heart knocked.

Upstairs, my 85-year-old father, Jeffrey, a retired Wesleyan University professor who suffered from dementia, lay napping in what

was once their shared bedroom. Sewn into a hump of skin and muscle below his right clavicle was the pacemaker that helped his heart outlive his brain. The size of a pocket watch, it had kept his heart beating rhythmically for nearly five years. Its battery was expected to last five more.

After tea, I knew, my mother would help him from his narrow bed with its mattress encased in waterproof plastic. She would take him to the toilet, change his diaper and lead him tottering to the couch, where he would sit mutely for hours, pretending to read Joyce Carol Oates, the book falling in his lap as he stared out the window.

I don't like describing what dementia did to my father—and indirectly to my mother—without telling you first that my parents loved each other, and I loved them. That my mother, Valerie, could stain a deck and sew an evening dress from a photo in *Vogue* and thought of my father as her best friend. That my father had never given up easily on anything.

Born in South Africa, he lost his left arm in World War II, but built floor-to-ceiling bookcases for our living room; earned a Ph.D. from Oxford; coached rugby; and with my two brothers as crew, sailed his beloved Rhodes 19 on Long Island Sound. When I was a child, he woke me, chortling, with his gloss on a verse from *The Rubaiyat of Omar Khayyam:* "Awake, my little one! Before life's liquor in its cup be dry!" At bedtime he tucked me in, quoting *Hamlet:* "May flights of angels sing thee to thy rest!"

Now I would look at him and think of Anton Chekhov, who died of tuberculosis in 1904. "Whenever there is someone in a family who has long been ill, and hopelessly ill," he wrote, "there come painful moments when all timidly, secretly, at the bottom of their hearts long for his death." A century later, my mother and I had come to long for the machine in my father's chest to fail.

Until 2001, my two brothers and I—all living in California—assumed that our parents would enjoy long, robust old ages capped by some brief, undefined final illness. Thanks to their own healthful habits and a panoply of medical advances—vaccines, antibiotics, airport defibrillators, 911 networks and the like—they weren't likely to die prematurely of the pneumonias, influenzas and heart attacks that decimated previous generations.

They walked every day. My mother practiced yoga. My father was writing a history of his birthplace, a small South African town.

In short, they were seemingly among the lucky ones for whom the American medical system, despite its fragmentation, inequity and waste, works quite well. Medicare and supplemental insurance paid for their specialists and their trusted Middletown internist, the lean, bespectacled Robert Fales, who, like them, was skeptical of medical overdoing.

"I bonded with your parents, and you don't bond with everybody," he once told me. "It's easier to understand someone if they just tell it like it is from their heart and their soul."

They were also stoics and religious agnostics. They signed living wills and durable power-of-attorney documents for health care. My mother, who watched friends die slowly of cancer, had an underlined copy of the Hemlock Society's *Final Exit* in her bookcase. Even so, I watched them lose control of their lives to a set of perverse financial incentives—for cardiologists, hospitals and especially the manufacturers of advanced medical devices—skewed to promote maximum treatment. At a point hard to precisely define, they stopped being beneficiaries of the war on sudden death and became its victims.

THINGS TOOK THEIR FIRST UNEXPECTED TURN on Nov. 13, 2001, when my father—then 79, pacemakerless and seemingly healthy—collapsed on my parents' kitchen floor in Middletown, making burbling sounds. He had suffered a stroke.

He came home six weeks later permanently incapable of completing a sentence. But as I've said, he didn't give up easily, and he doggedly learned again how to fasten his belt; to peck out sentences on his computer; to walk alone, one foot dragging, to the university pool for water aerobics. He never again put on a shirt without help or looked at the book he had been writing. One day he haltingly told my mother, "I don't know who I am anymore."

His stroke devastated two lives. The day before, my mother was an upper-middle-class housewife who practiced calligraphy in her spare time. Afterward, she was one of tens of millions of people in America, most of them women, who help care for an older family member.

Their numbers grow each day. Thanks to advanced medical technologies, elderly people now survive repeated health crises that once killed them, and so the "oldest old" have become the nation's most rapidly growing age group. Nearly a third of Americans over 85 have dementia (a condition whose prevalence rises in direct relationship to longevity). Half need help with at least one practical, life-sustaining activity, like getting dressed or making breakfast. Even though a capable woman was hired to give my dad showers, my 77-year-old mother found herself on duty more than 80 hours a week. Her blood pressure rose and her weight fell. On a routine visit to Dr. Fales, she burst into tears. She was put on sleeping pills and antidepressants.

My father said he came to believe that she would have been better off if he had died. "She'd have weeped the weep of a widow," he told me in his garbled, poststroke speech, on a walk we took together in the fall of 2002. "And then she would have been all right." It was hard to tell which of them was suffering more.

As we shuffled through the fallen leaves that day, I thought of my father's father, Ernest Butler. He was 79 when he died in 1965, before pacemakers, implanted cardiac defibrillators, stents and replacement heart valves routinely staved off death among the very old. After completing some long-unfinished chairs, he cleaned his woodshop, had a heart attack and died two days later in a plain hospital bed. As I held my dad's soft, mottled hand, I vainly wished him a similar merciful death.

A few days before Christmas that year, after a vigorous session of water exercises, my father developed a painful inguinal (intestinal) hernia. My mother took him to Fales, who sent them to a local surgeon, who sent them to a cardiologist for a preoperative clearance. After an electrocardiogram recorded my father's slow heartbeat—a longstanding and symptomless condition not uncommon in the very old—the cardiologist, John Rogan, refused to clear my dad for surgery unless he received a pacemaker.

Without the device, Dr. Rogan told me later, my father could have died from cardiac arrest during surgery or perhaps within a few months. It was the second time Rogan had seen my father. The first time, about a year before, he recommended the device for the same slow heartbeat. That time, my then-competent and prestroke father expressed extreme reluctance, on the advice of Fales, who considered it overtreatment.

My father's medical conservatism, I have since learned, is not unusual. According to an analysis by the Dartmouth Atlas medical-research group, patients are far more likely than their doctors to reject aggressive treatments when fully informed of pros, cons and alternatives—information, one study suggests, that nearly half of patients say they don't get. And although many doctors assume that people want to extend their lives, many do not. In a 1997 study in *The Journal of the American Geriatrics Society,* 30 percent of seriously ill people surveyed in a hospital said they would "rather die" than live permanently in a nursing home. In a 2008 study in *The Journal of the American College of Cardiology,* 28 percent of patients with advanced heart failure said they would trade one day of excellent health for another two years in their current state.

When Rogan suggested the pacemaker for the second time, my father was too stroke-damaged to discuss, and perhaps even to weigh, his tradeoffs. The decision fell to my mother—anxious to relieve my father's pain, exhausted with caregiving, deferential to doctors and no expert on high-tech medicine. She said yes. One of the most important medical decisions of my father's life was over in minutes. Dr. Fales was notified by fax.

FALES LOVED MY PARENTS, KNEW THEIR suffering close at hand, continued to oppose a pacemaker and wasn't alarmed by death. If he had had the chance to sit down with my parents, he could have explained that the pacemaker's battery would last 10 years and asked whether my father wanted to live to be 89 in his nearly mute and dependent state. He could have discussed the option of using a temporary external pacemaker that, I later learned, could have seen my dad safely through surgery. But my mother never consulted Fales. And the system would have effectively penalized him if she had. Medicare would have paid him a standard office-visit rate of $54 for what would undoubtedly have been a long meeting—and nothing for phone calls to work out a plan with Rogan and the surgeon.

Medicare has made minor improvements since then, and in the House version of the health care reform bill debated last year, much better payments for such conversations were included. But after the provision was distorted as reimbursement for "death panels," it was dropped. In my father's case, there was only a brief informed-consent process, covering the boilerplate risks of minor surgery, handled by the general surgeon.

I believe that my father's doctors did their best within a compartmentalized and time-pressured medical system. But in the absence of any other guiding hand, there is no doubt that economics helped shape the wider context in which doctors made decisions. Had we been at the Mayo Clinic—where doctors are salaried, medical records are electronically organized and care is coordinated by a single doctor—things might have turned out differently. But Middletown is part of the fee-for-service medical economy. Doctors peddle their wares on a piecework basis; communication among them is haphazard; thinking is often short term; nobody makes money when medical interventions are declined; and nobody is in charge except the marketplace.

And so on Jan. 2, 2003, at Middlesex Hospital, the surgeon implanted my father's pacemaker using local anesthetic. Medicare paid him $461 and the hospital a flat fee of about $12,000, of which an estimated $7,500 went to St. Jude Medical, the maker of the device. The hernia was fixed a few days later.

It was a case study in what primary-care doctors have long bemoaned: that Medicare rewards doctors far better for doing procedures than for assessing whether they should be done at all. The incentives for overtreatment continue, said Dr. Ted Epperly, the board chairman of the American Academy of Family Physicians, because those who profit from them—specialists, hospitals, drug companies and the medical-device manufacturers—spend money lobbying Congress and the public to keep it that way.

Last year, doctors, hospitals, drug companies, medical-equipment manufacturers and other medical professionals spent $545 million on lobbying, according to the Center for Responsive Politics. This may help explain why researchers estimate that 20 to 30 percent of Medicare's $510 billion budget goes for unnecessary tests and treatment. Why cost-containment received short shrift in health care reform. Why physicians like Fales net an average of $173,000 a year, while noninvasive cardiologists like Rogan net about $419,000.

The system rewarded nobody for saying "no" or even "wait"—not even my frugal, intelligent, *Consumer-Reports*-reading mother. Medicare and supplemental insurance covered almost every penny of my father's pacemaker. My mother was given more government-mandated consumer information when she bought a new Camry a year later.

And so my father's electronically managed heart—now requiring frequent monitoring, paid by Medicare—became part of the $24 billion worldwide cardiac-device industry and an indirect subsidizer of the fiscal health of American hospitals. The profit margins that manufacturers earn on cardiac devices is close to 30 percent. Cardiac procedures and diagnostics generate about 20 percent of hospital revenues and 30 percent of profits.

* * *

SHORTLY AFTER NEW YEAR'S 2003, MY mother belatedly called and told me about the operations, which went off without a hitch. She didn't call earlier, she said, because she didn't want to worry me. My heart sank, but I said nothing. It is one thing to silently hope that your beloved father's heart might fail. It is another to actively abet his death.

The pacemaker bought my parents two years of limbo, two of purgatory and two of hell. At first they soldiered on, with my father no better and no worse. My mother reread Jon Kabat-Zinn's *Full Catastrophe Living*, bought a self-help book on patience and rose each morning to meditate.

In 2005, the age-related degeneration that had slowed my father's heart attacked his eyes, lungs, bladder and bowels. Clots as narrow as a single human hair lodged in tiny blood vessels in his brain, killing clusters of neurons by depriving them of oxygen. Long partly deaf, he began losing his sight to wet macular degeneration, requiring ocular injections that cost nearly $2,000 each. A few months later, he forgot his way home from the university pool. He grew incontinent. He was collapsing physically, like an ancient, shored-up house.

In the summer of 2006, he fell in the driveway and suffered a brain hemorrhage. Not long afterward, he spent a full weekend compulsively brushing and rebrushing his teeth. "The Jeff I married . . . is no longer the same person," my mother wrote in the journal a social worker had suggested she keep. "My life is in ruins. This is horrible, and I have lasted for five years." His pacemaker kept on ticking.

When bioethicists debate life-extending technologies, the effects on people like my mother rarely enter the calculus. But a 2007 Ohio State University study of the DNA of family caregivers of people with Alzheimer's disease showed that the ends of their chromosomes, called telomeres, had degraded enough to reflect a four-to-eight-year shortening of lifespan. By that reckoning, every year that the pacemaker gave my irreparably damaged father took from my then-vigorous mother an equal year.

When my mother was upset, she meditated or cleaned house. When I was upset, I Googled. In 2006, I discovered that pacemakers could be deactivated without surgery. Nurses, doctors and even device salesmen had done so, usually at deathbeds. A white ceramic device, like a TV remote and shaped like the wands that children use to blow bubbles, could be placed around the hump on my father's chest. Press a few buttons and the electrical pulses that ran down the leads to his heart would slow until they were no longer effective. My father's heart, I learned, would probably not stop. It would just return to its old, slow rhythm. If he was lucky, he might suffer cardiac arrest and die within weeks, perhaps in his sleep. If he was unlucky, he might linger painfully for months while his lagging heart failed to suffuse his vital organs with sufficient oxygenated blood.

If we did nothing, his pacemaker would not stop for years. Like the tireless charmed brooms in Disney's *Fantasia,* it would prompt my father's heart to beat after he became too demented to speak, sit up or eat. It would keep his heart pulsing after he drew his last breath. If he was buried, it would send signals to his dead heart in the coffin. If he was cremated, it would have to be cut from his chest first, to prevent it from exploding and damaging the walls or hurting an attendant.

ON THE INTERNET, I DISCOVERED that the pacemaker— somewhat like the ventilator, defibrillator and feeding tube—was first an exotic, stopgap device, used to carry a handful of patients through a brief medical crisis. Then it morphed into a battery-powered, implantable and routine treatment. When Medicare approved the pacemaker for reimbursement in 1966, the market exploded. Today pacemakers are implanted annually in more than 400,000 Americans, about 80 percent of whom are over 65. According to calculations by the Dartmouth Atlas research group using Medicare data, nearly a fifth of new recipients who receive pacemakers annually—76,000—are over 80. The typical patient with a cardiac device today is an elderly person suffering from at least one other severe chronic illness.

Over the years, as technology has improved, the battery life of these devices lengthened. The list of heart conditions for which they are recommended has grown. In 1984, the treatment guidelines from the American College of Cardiology declared that pacemakers were strongly recommended as "indicated" or mildly approved as "reasonable" for 56 heart conditions and "not indicated" for 31 more. By 2008, the list for which they were strongly or mildly recommended expanded to 88, with most of the increase in the lukewarm "reasonable" category.

The research backing the expansion of diagnoses was weak. Over all, only 5 percent of the positive recommendations were supported by research from multiple double-blind randomized studies, the gold standard of evidence-based medicine. And 58 percent were based on no studies at all, only a "consensus of expert opinion." Of the 17 cardiologists who wrote the 2008 guidelines, 11 received financing from cardiac-device makers or worked at institutions receiving it. Seven, due to the extent of their financial connections, were recused from voting on the guidelines they helped write.

This pattern—a paucity of scientific support and a plethora of industry connections—holds across almost all cardiac treatments, according to the cardiologist Pierluigi Tricoci of Duke University's Clinical Research Institute. Last year in *The Journal of the American*

Medical Association, Tricoci and his co-authors wrote that only 11 percent of 2,700 widely used cardiac-treatment guidelines were based on that gold standard. Most were based only on expert opinion.

Experts are as vulnerable to conflicts of interest as researchers are, the authors warned, because "expert clinicians are also those who are likely to receive honoraria, speakers bureau [fees], consulting fees or research support from industry." They called the current cardiac-research agenda "strongly influenced by industry's natural desire to introduce new products."

Perhaps it's no surprise that I also discovered others puzzling over cardiologists who recommended pacemakers for relatives with advanced dementia. "78-year-old mother-in-law has dementia; severe short-term memory issues," read an Internet post by "soninlaw" on Elderhope.com, a caregivers' site, in 2007. "On a routine trip to her cardiologist, doctor decides she needs a pacemaker. . . . Anyone have a similar encounter?"

By the summer of 2007, my dad had forgotten the purpose of a dinner napkin and had to be coached to remove his slippers before he tried to put on his shoes. After a lifetime of promoting my father's health, my mother reversed course. On a routine visit, she asked Rogan to deactivate the pacemaker. "It was hard," she later told me. "I was doing for Jeff what I would have wanted Jeff to do for me." Rogan soon made it clear he was morally opposed. "It would have been like putting a pillow over your father's head," he later told me.

Not long afterward, my mother declined additional medical tests and refused to put my father on a new anti-dementia drug and a blood thinner with troublesome side effects. "I take responsibility for whatever," she wrote in her journal that summer. "Enough of all this overkill! It's killing me! Talk about quality of life—what about mine?"

THEN CAME THE AUTUMN DAY when she asked for my help, and I said yes. I told myself that we were simply trying to undo a terrible medical mistake. I reminded myself that my dad had rejected a pacemaker when his faculties were intact. I imagined, as a bioethicist had suggested, having a 15-minute conversation with my independent, predementia father in which I saw him shaking his head in horror over any further extension of what was not a "life," but a prolonged and attenuated dying. None of it helped. I knew that once he died, I would dream of him and miss his mute, loving smiles. I wanted to melt into the arms of the father I once had and ask him to handle this. Instead, I felt as if I were signing on as his executioner and that I had no choice.

Over the next five months, my mother and I learned many things. We were told, by the Hemlock Society's successor, Compassion and Choices, that as my father's medical proxy, my mother had the legal right to ask for the withdrawal of any treatment and that the pacemaker was, in theory at least, a form of medical treatment. We learned that although my father's living will requested no life support if he were comatose or dying, it said nothing about dementia and did not define a pacemaker as life support. We learned that if we called 911, emergency medical technicians would not honor my father's do-not-resuscitate order unless he wore a state-issued orange hospital bracelet. We also learned that no cardiology association had given its members clear guidance on when, or whether, deactivating pacemakers was ethical.

(Last month that changed. The Heart Rhythm Society and the American Heart Association issued guidelines declaring that patients or their legal surrogates have the moral and legal right to request the withdrawal of any medical treatment, including an implanted cardiac device. It said that deactivating a pacemaker was neither euthanasia nor assisted suicide, and that a doctor could not be compelled to do so in violation of his moral values. In such cases, it continued, doctors "cannot abandon the patient but should involve a colleague who is willing to carry out the procedure." This came, of course, too late for us.)

In the spring of 2008, things got even worse. My father took to roaring like a lion at his caregivers. At home in California, I searched the Internet for a sympathetic cardiologist and a caregiver to put my dad to bed at night. My frayed mother began to shout at him, and their nighttime scenes were heartbreaking and frightening. An Alzheimer's Association support-group leader suggested that my brothers and I fly out together and institutionalize my father. This leader did not know my mother's formidable will and had never heard her speak about her wedding vows or her love.

Meanwhile my father drifted into what nurses call "the dwindles": not sick enough to qualify for hospice care, but sick enough to never get better. He fell repeatedly at night and my mother could not pick him up. Finally, he was weak enough to qualify for palliative care, and a team of nurses and social workers visited the house. His chest grew wheezy. My mother did not request antibiotics. In mid-April 2008, he was taken by ambulance to Middlesex Hospital's hospice wing, suffering from pneumonia.

Pneumonia was once called "the old man's friend" for its promise of an easy death. That's not what I saw when I flew in. On morphine, unreachable, his eyes shut, my beloved father was breathing as hard and regularly as a machine.

My mother sat holding his hand, weeping and begging for forgiveness for her impatience. She sat by him in agony. She beseeched his doctors and nurses to increase his morphine dose and to turn off the pacemaker. It was a weekend, and the doctor on call at Rogan's cardiology practice refused authorization, saying that my father "might die immediately." And so came five days of hard labor. My mother and I stayed by him in shifts, while his breathing became increasingly ragged and his feet slowly started to turn blue. I began drafting an appeal to the hospital ethics committee. My brothers flew in.

On a Tuesday afternoon, with my mother at his side, my father stopped breathing. A hospice nurse hung a blue light on the outside of his hospital door. Inside his chest, his pacemaker was still quietly pulsing.

After his memorial service in the Wesleyan University chapel, I carried a box from the crematory into the woods of an old convent where he and I often walked. It was late April, overcast and cold. By the side of a stream, I opened the box, scooped out a handful of ashes and threw them into the swirling water. There were some curious spiraled metal wires, perhaps the leads of his pacemaker, mixed with the white dust and pieces of bone.

A YEAR LATER, I TOOK my mother to meet a heart surgeon in a windowless treatment room at Brigham and Women's Hospital in Boston. She was 84, with two leaking heart valves. Her cardiologist had recommended open-heart surgery, and I was hoping to find a less invasive approach. When the surgeon asked us why we were there, my mother said, "To ask questions." She was no longer a trusting and deferential patient. Like me, she no longer saw doctors—perhaps with the exception of Fales—as healers or her fiduciaries. They were now skilled technicians with their own agendas. But I couldn't help feeling that something precious—our old faith in a doctor's calling, perhaps, or in a healing that is more than a financial transaction or a reflexive fixing of broken parts—had been lost.

The surgeon was forthright: without open-heart surgery, there was a 50-50 chance my mother would die within two years. If she survived the operation, she would probably live to be 90. And the risks? He shrugged. Months of recovery. A 5 percent chance of stroke. Some possibility, he acknowledged at my prompting, of post-operative cognitive decline. (More than half of heart-bypass patients suffer at least a 20 percent reduction in mental function.) My mother lifted her trouser leg to reveal an anklet of orange plastic: her do-not-resuscitate bracelet. The doctor recoiled. No, he would not operate with that bracelet in place. It would not be fair to his team. She would be revived if she collapsed.

"If I have a stroke," my mother said, nearly in tears, "I want you to let me go." What about a minor stroke, he said—a little weakness on one side?

I kept my mouth shut. I was there to get her the information she needed and to support whatever decision she made. If she emerged from surgery intellectually damaged, I would bring her to a nursing home in California and try to care for her the way she had cared for my father at such cost to her own health. The thought terrified me.

The doctor sent her up a floor for an echocardiogram. A half-hour later, my mother came back to the waiting room and put on her black coat. "No," she said brightly, with the clarity of purpose she had shown when she asked me to have the pacemaker deactivated. "I will not do it."

She spent the spring and summer arranging house repairs, thinning out my father's bookcases and throwing out the files he collected so lovingly for the book he never finished writing. She told someone that she didn't want to leave a mess for her kids. Her chest pain worsened, and her breathlessness grew severe. "I'm aching to garden," she wrote in her journal. "But so it goes. ACCEPT ACCEPT ACCEPT."

Last August, she had a heart attack and returned home under hospice care. One evening a month later, another heart attack. One of my brothers followed her ambulance to the hospice wing where we had sat for days by my father's bed. The next morning, she took off her silver earrings and told the nurses she wanted to stop eating and drinking, that she wanted to die and never go home. Death came to her an hour later, while my brother was on the phone to me in California—almost as mercifully as it had come to my paternal grandfather. She was continent and lucid to her end.

A week later, at the same crematory near Long Island Sound, my brothers and I watched through a plate-glass window as a cardboard box containing her body, dressed in a scarlet silk *ao dai* she had sewn herself, slid into the flames. The next day, the undertaker delivered a plastic box to the house where, for 45 of their 61 years together, my parents had loved and looked after each other, humanly and imperfectly. There were no bits of metal mixed with the fine white powder and the small pieces of her bones.

JOHN COLAPINTO

Mother Courage

FROM *THE NEW YORKER*

A rare and lethal form of muscular dystrophy diagnosed in her sons turned one mother into an advocate who has helped raise hundreds of millions of dollars in research—even though, as John Colapinto reports, it was too late for her children.

IN THE EARLY NINETEEN-EIGHTIES, PAT FURLONG noticed that her infant son, Patrick, was "floppy." He would slip through her hands when she attempted to lift him from under his armpits. "You'd stand him up," Furlong says, "and he'd just kind of blob down." Her two daughters, then seven and five, were developing normally, but she began to look with concern at her other son, Christopher, who was two years older than Patrick. He, too, had been delayed in his motor skills; he had walked a little later than other kids, and now, at three years old, he could not turn a somer-sault. Furlong took the boys to doctors all over their home town, Middletown, Ohio. She was told that there was nothing wrong with them. Both boys were smiling and engaging, clearly normal in their mental functioning. But Furlong, a health educator and a former nurse, remained worried. "For me, it was a feeling in my stomach that something was wrong," she says. She complained to her husband, Tom, a family-practice physician, that other children in the neighborhood ran and jumped while Chris and Pat sat coloring. Tom dismissed her fears, saying that the other kids were simply unruly.

As a nurse, Furlong had spent years examining illness. She grew up in Cincinnati, the youngest child of first-generation German-American parents, and in 1969 she earned a nursing degree from Mount St. Joseph's, a local college. "I'm fascinated by medicine— what we can do and can't do," she told me. In graduate school, at Ohio State, she ran the intensive-care unit at the campus hospital. But her experience caring for adult patients was of no help in trying to understand what was wrong with the boys. "If an adult was sick, I could go through it systematically and say, 'What are we looking at?'" she said. "But with children I didn't have those tools."

When Chris was six, he suffered an unexplained injury to his calf. "He was riding his Big Wheel tricycle," Furlong told me. "Pointing his toe and flexing, he said all of a sudden that something hurt. And he started to cry. It was weird, because what could he have done? His calf muscle started to really swell." Enlarged calves are a leading di-agnostic indicator of Duchenne muscular dystrophy, a rapid, fatal muscle-wasting disease that affects males almost exclusively. As the muscles die, they are replaced by scar tissue that looks, to the unini-tiated, like increased muscle mass. Furlong had noted the unusual bulk of her sons' calves, but had taken it as an indication that the boys would grow up to be big, like their father, who had played foot-ball for Notre Dame.

Tom seemed mystified by Christopher's injury. But the next morning Furlong saw him looking at Chris with tears in his eyes. "I said, 'What are you looking at?'" Furlong recalls. "And he said, 'I don't know.'" Later, Tom admitted that the possibility of Duchenne had crossed his mind. "But it's not a diagnosis you want to make," Pat says. "And there was no family history of the disease."

Still hoping that the injury was a sprain or a muscle tear, Furlong made an appointment with an orthopedic surgeon in Middletown. The doctor saw Chris's calves and his distinctive gait—he lifted his shoulders and pushed his pelvis forward, to help swing his weakened legs—and knew immediately that the boy had Duchenne. "He didn't even have to do any tests," Furlong says. "He said, 'What does Patrick look like?' And I said, 'Well, he's the same.'"

Furlong was referred to Cincinnati Children's Hospital, where the boys underwent three days of testing. A neurologist there confirmed the diagnosis of Duchenne, and laid out a dire prognosis. The first signs of the disease usually appear in early childhood and are followed by rapid, progressive loss of muscle strength that lands sufferers in a wheelchair by their early teens and eventually renders them completely immobile. Victims typically die from cardiac or respiratory failure, often before the age of twenty. Since 1861, when the French neurologist Guillaume Duchenne first described the disease, no drug had been developed that affected its inexorable course.

Furlong recalls that the neurologist told her, "There is no hope and no help—just take them home and love them. They're going to die." Duchenne is a genetic condition, coded on the mother's X chromosome, and is usually inherited. The doctor upbraided her, she says, for having had a second boy. "'You should have known about this,'" she recalls him telling her. "'This is a familial disease, it's genetic, you have it in your family.' I said, 'I don't.'" (She learned later that she was among the one-third of cases in which the mutation appears spontaneously.) The doctor insisted, "You could have prevented the second pregnancy, or you could have aborted the second pregnancy.'" Patrick, then four, was sitting on Furlong's lap. "Before that day, I was relatively mild-mannered," Furlong says. She remembers grabbing the doctor's tie and pulling him up to her nose. "I said, 'If somebody should have been aborted today, you're the one.'"

Furlong was determined to fight the disease, but Tom, as a physician, tended to accept the prevailing medical view that nothing could be done. He began to prepare for when Christopher and Patrick could no longer walk, building a wheelchair-accessible addition to their house and installing a small elevator. Furlong was furious: "I wasn't where I am now, able to stand back and say, 'He was crushed in his own way, and he was trying to survive.'" Tom withdrew from the family, shutting himself in his home office in the evenings and refusing to discuss the diagnosis with anyone. The two grew apart, as many Duchenne parents do.

As a nurse, Furlong understood that her sons would probably die of Duchenne. Still, she believed that she could somehow save them. "In my head," she says, "I could get there in time."

DUCHENNE—THE DEADLIEST OF THE MORE than forty disorders that go under the name muscular dystrophy—occurs once in every thirty-five hundred live male births. There are about two hundred and fifty thousand sufferers worldwide, between ten and twenty thousand of them in the United States. Duchenne is classified by the World Health Organization as a rare disease, and scientists often see little incentive to specialize in such diseases, because few research dollars are available. Thus, the burden of stimulating new research often falls on patients or their parents.

Furlong, at sixty-four, is a tall, attractive woman with a dramatic presentation—bright-red lipstick, ruffled black tops, artfully mussed brown hair—and a manner that blends unyielding resolve with self-deprecating humor and a certain sadness. She has been singularly effective as a parent activist, not only in spurring research into a cure for the disease but in working with doctors and drug companies to improve care. In 2001, she helped lobby Congress to pass legislation that has allocated more than four hundred million dollars for research into muscular dystrophy, of which more than a hundred and sixty million has gone to Duchenne research—far outstripping the research money raised for Duchenne by the Muscular Dystrophy Association. Dr. Steve Groft, the director of the Office of Rare Diseases Research, at the National Institutes of Health, told me recently that much of the current clinical research into the disease is the result of Furlong's efforts: "She has been the major mover with Duchenne muscular dystrophy—around the world."

LESS THAN A WEEK AFTER her sons received their diagnosis, in June, 1984, Furlong went to the bank and, without her husband's knowledge, borrowed a hundred thousand dollars. "The president of the bank said, 'Does Tom know?'" Furlong recalled. "I said, 'Of course—would I be here if he didn't?'" She forged his signature as co-signer on the loan. Tom found out and "went wild," she says. When she told him that she was going to use the money to save their sons, he told her that it was a futile effort, but he did not force her to return the money. Furlong says he knew that "if he messed with me he would lose me."

Furlong began travelling to medical centers around the United States and Europe. "I wanted to understand the landscape—who the players were, what they thought, how they thought about Duchenne, what priorities, what plans," she says. She found that very little was being done, and that few doctors were willing to talk to her. "Most researchers and physicians will do anything to avoid meeting with distraught mothers," she says. To get past secretaries, she would impersonate a doctor on the phone, then arrive at meetings dressed for the part. "I would wear something very professional and the highest heels possible," she said. She added, "Of course, I'd start crying the minute I sat down with them—so they knew."

Not long into her research, she read a paper by Dr. Charles Bonsett, a neurologist in Indiana, who reported promising early results from adenylosuccinate acid. She flew to Indiana to meet him. No clinical trials for Duchenne existed, but Furlong persuaded Bonsett to launch one using the compound. She helped fund it with the money she had borrowed from the bank, and made sure that her sons were included. They took the drug—which was administered by a shunt surgically implanted in their abdomens—for two years, but their slide continued. At eight, Christopher lost the ability to walk; the same year, Patrick, whose Duchenne was more aggressive, lost the use of his legs, too. Subsequently, they enrolled in a treatment, overseen by a doctor in Memphis, that involved the transplantation of immature muscle cells, called myoblasts. It was no more successful, and Furlong withdrew them from the trial in 1993.

By then, doctors had discovered that steroids can slow muscle wasting in Duchenne boys, extending life into the twenties and sometimes beyond. (The drugs have serious side effects, including severe mood alteration, weight gain, stunted growth, suppressed puberty, and increased bone fragility.) Clinical trials of steroids for Duchenne were in progress, and Furlong tried to get her sons included. But, when she was told that a placebo group would not be given the treatment, she withdrew them, rather than take the risk that one of them would not receive the drug. Instead, she asked Tom to write a prescription for steroids. He refused, saying that they could not perform unregulated drug experiments on their children. Furlong, incensed, forged his handwriting on a prescription, but finally decided not to fill it. "I did not want to be culpable," she said.

At the ages of fifteen and thirteen, Christopher and Patrick were fully paralyzed, except for their organs of speech and their fingers. They moved around in motorized wheelchairs using joysticks, and relied on family members and a few paid helpers to lift them onto the toilet, or from wheelchair to bed. Unable to raise their arms, they ate with their elbows propped on high drafting tables that their father rigged for them. Despite these difficulties, Christopher regularly ranked at the top of his class in school; Patrick concentrated on friends and dreamed of success as a gambler. When the myoblast treatment failed, they were crestfallen. "Chris asked, 'Have you given up on us?'" Furlong recalls. She assured him that she had not, and shortly afterward, in 1994, she decided to start a patient-advocacy group.

"By this time, I was well acquainted with the small group of researchers in the field," Furlong says. "I thought it was time to call them together with the intention of funding research. To raise significant dollars, we needed to form a nonprofit." She got in touch with Duchenne parents whom she had met through clinical trials, and incorporated the organization under the name Muscular Dystrophy Research Foundation. Immediately, she received a letter from the executive director of the Muscular Dystrophy Association, Robert Ross, accusing her of deliberately choosing a name that would confuse the public. "He was wrong," she says. "I just wasn't very creative." She changed the name to Parent Project Muscular Dystrophy.

The twenty or so parents who made up the group held bake sales and dinner dances and canvassed family, friends, and neighbors for donations. "For one holiday season, we were baking and sending out holiday cards to our donors," Furlong says. "I devised this holiday card full of sparkles and set up an assembly line in my kitchen, someone adding the glue to the card, the next person adding the sparkles." Her motivation in starting the parent group was not sadness but anger, she told me: "at God, at the world, at being in Middletown, Ohio, not having the academic resources at my fingertips, having doctors who say I'm crazy, angry at a husband who doesn't seem to be in synch with me, angry at myself for a genetic disease."

* * *

IN 1986, LOUIS KUNKEL, A research scientist at Boston University, isolated the gene implicated in Duchenne. This gene—by far the largest in the genome, with 2.3 million base pairs—is responsible for making a protein that Kunkel and his collaborators named dystrophin. The protein is found in every muscle of the body; it forms a shock absorber around the membrane that surrounds muscle cells and holds them together. Owing to errors in their genetic code, Duchenne boys do not produce dystrophin. Over time, the cell membranes rip, and calcium floods into the cells. The immune system, mistaking the activity for an invading pathogen, attacks the cells and kills them. Muscles fail progressively throughout the body: first the large skeletal muscles of the legs and arms and trunk, then the muscles of the diaphragm, digestive tract, and heart.

In the early nineties, Furlong began visiting gene-therapy researchers, including Eric Hoffman, a geneticist who had worked in Kunkel's lab and then established an independent laboratory at the University of Pittsburgh. Furlong asked Hoffman to address the members of Parent Project, whom she was bringing together for an information-sharing meeting in Ontario, in the spring of 1994. Furlong says that many researchers were leery of associating with Parent Project out of fear, "real or imagined," that the Muscular Dystrophy Association would cancel their funding. But, she says, "if Eric Hoffman agreed to come, we knew the rest would come along."

Hoffman, however, declined the invitation. "He said, 'No one's going to come,'" Furlong recalls. "'What would incentivize us?'" On a table in Hoffman's office was a copy of *Time*, with a cover story on the geneticist French Anderson, who had performed the world's first successful gene therapy, on children with ADA deficiency, an immune disorder. In desperation, Furlong grabbed the magazine and said, "He's coming." Hoffman was clearly dubious. "He just looked at me, like, 'You're the dumbest shit I've ever come across.'"

Furlong immediately flew to Pasadena, where Anderson had an office at U.S.C., and installed herself in his waiting room until he emerged. "I told him the whole story," she says. "I'm in tears: 'You have to do this, I made up a big lie, oh, my gosh—I'm crazy, I'm a liar, but you have to come!' He just said, 'O.K.'" Furlong shrugged. "I think he was afraid of me."

At the meeting in Ontario, Anderson (who, in 2006, was convicted of sexual abuse of a minor and is currently serving a fourteen-year prison sentence) announced to a crowd of parents, patients, and scientists that he would cure Duchenne in eighteen months—a grandiose claim that almost no one in the room believed. But Furlong had managed to assemble several of the world's leading experts on the disease, including Hoffman. "The meeting galvanized everyone," Furlong says. "I recall talking to the researchers after the meeting. They were pretty amazed that parents were not hysterical, did not expect magic, but wanted to do whatever they could to help. They asked questions about the disease and about us, exploring ideas."

The parents began raising money to establish a research center. They agreed that the University of Pittsburgh was the logical place; Hoffman was there, and it had a number of "bright minds in muscle and gene therapy," Furlong says. The parents pledged to raise three hundred thousand dollars. "We had about one hundred and fifty thousand in our accounts," she says. In the next twelve months, the group raised the rest—through more bake sales, "dinner dances, letters to friends and relatives, and general begging"—and the Duchenne Muscular Dystrophy Research Center, the first of its kind, opened in 1995. Furlong says, "We found a place to land—a place to focus efforts, a place where we no longer felt like 'just parents.'"

LEE SWEENEY, THE SCIENTIFIC ADVISER to Parent Project Muscular Dystrophy, has been studying the molecular basis of muscle movement since 1984, and is now the chairman of the Department of Physiology at the University of Pennsylvania School of Medicine. He was engaged in gene experiments in Duchenne using mice when, in the mid-nineties, Furlong sought him out and offered funding. "Parents, the minute they find the disease, do one of several things," Sweeney told me. "Either they withdraw and just feel like the world has attacked them or they decide that they're going to do something about it, and attack back. Often they do what Pat did, early on, which is to try to form some charity—usually in the name of the child—and they go out and raise money from their friends and other people." But developing a drug from initial lab experiments to F.D.A. approval costs hundreds of millions of dollars—far more than grass-roots charities can raise. "When I first met Pat, I thought, It's another one of these people, and I feel badly for her," he said. He turned down her offer. "The type of money she was talking about was relatively small," he said, and the gene approaches he was working on were far from yielding any benefit to humans. But Sweeney agreed to speak at an annual meeting of Parent Project Muscular Dystrophy.

In October 1995, the Furlongs' older son, Christopher, then seventeen, caught a cold that developed rapidly into pneumonia—a common occurrence in Duchenne boys, whose weak diaphragms make it impossible for them to clear their lungs by coughing. (Furlong has since worked with the pediatric pulmonologist Jonathan Finder to include the mechanical insufflator-exsufflator—a device that aids in coughing—in the standard care for Duchenne.) Christopher was admitted to the hospital, intubated, and placed on a respirator. His condition worsened, and, after two weeks, he died.

Seven months later, Patrick, too, got a cold, and his health declined. "After Chris died, I think Patrick gave up," Furlong told me. "He watched his brother lose strength and die. He knew. Patrick and I could barely look eye to eye without tears. It was as if we knew a secret and if one or the other said it out loud it would come true."

Furlong took Patrick to the hospital and stayed with him all night. In the morning, he said that she should go home and shower, and asked that she get his father to spell her. He joked that "having a doctor around was much better than a nurse." Furlong summoned Tom, then drove home. She was at the door of their house when Tom called her cell phone to tell her that Patrick's lungs and heart had failed.

Few parents, Sweeney says, remain committed to fighting Duchenne after their child's death: "I thought that, like most of the parents, Pat would never want to hear the name of the disease again—she would disappear, her organization would disappear, and that would be it. But, to my surprise, if anything she came back to me even more aggressive." Furlong says that she never contemplated quitting. "I don't understand how you can close the door and say, 'This part of my life didn't exist,'" she told me. She was also inspired by something that Christopher said to her shortly after he and his brother first received their diagnosis of Duchenne. "We were in the kitchen and I was getting ready for dinner," Furlong said. "I must have been crying. Chris said to me, 'Why are you upset?' and I said, 'I just want you two cured.' And he said, 'Do you think it's fair if it's just Pat and me?' Just in a very weird kind of kidlike way. 'Wouldn't you want everyone cured? Aren't there more kids like this?'"

ONE OF SWEENEY'S FIRST ACTS as scientific adviser to Parent Project was to speak frankly to its members about the ineffectiveness of their approach. "At a meeting, I told them, just flat out, 'You know, this raising money from your friends and from bake sales is never going to get you research dollars that are going to have any impact. Maybe it makes you feel better, but it's not going to do anything.' My advice to them was to use the power of parents lobbying for children—which is a pretty powerful message—with their congressmen."

Furlong had been trying for years to make headway in Washington. "Right after the boys were diagnosed, I was at the N.I.H. asking, 'What are you doing?'" Furlong told me. "The answer was nothing." She was introduced to Steve Groft, at the Office of Rare Diseases Research, and began visiting his office, which she describes as "Mother Teresa's waiting room of rare diseases." Many concerned parents came to ask for help, she says, "but I was a pretty persistent one, going back, and calling Steve and saying, 'What are we doing here?' I am, by nature, a pest."

Furlong's unusual knowledge of the disease made her persuasive, Groft said. "With her training in nursing, she had a very good understanding of diseases and the needs of patients and families, and, as a mother of patients, she knew very well what her needs were." But his office by itself could be of limited help. "Our office was not very large," Groft explained. "I think we had two people." Following Groft's advice, Furlong met with the directors of several divisions of the N.I.H. that were potentially relevant to Duchenne, and with Dr. Harold Varmus, the director of the agency. Furlong recalled that in one meeting someone said, "We have heard about you, Mrs. Furlong." She considered that a victory. In 2000, after six years of appeals, she received a commitment from the N.I.H. to fund a "workshop" in Duchenne—a two-day conclave of researchers whose areas of focus were related to the disease. But when she told Sweeney about the workshop he was unimpressed. "He said, 'They're going to give you a workshop because you're in the queue,'" Furlong told me. "'Here's your workshop, now go away. It's not enough.' I realized he was right."

Furlong had recently met a Washington lobbyist named Joel Wood, who had a son with Duchenne. Wood put her in touch with B&D Consulting, a lobbying firm with expertise in pediatric legislation. Furlong and others from Parent Project worked with the lobbyists on a proposed bill that would compel the N.I.H. to fund a number of research centers for muscular dystrophy, with two devoted to Duchenne. Furlong then spent months visiting the office of Bettilou Taylor, the staff director of the Senate subcommittee in charge of health appropriations, to press for a hearing. Furlong describes Taylor as a formidable figure: "She was in her St. John's perfect outfits and the hair perfect and the makeup perfect and the Nancy Pelosi perfect face with the perfect smile and the sort of 'Tell me your story. Yes, I understand, I care a lot—next!'" Taylor was unable to schedule a hearing. Then, one day in late 2000, Furlong arrived with another Duchenne mother. Before they met with Taylor, Furlong told her not to sob; they had to be professional. "We went into this little conference room," Furlong recalls, "and she put her arms around Bettilou—put her tears on Bettilou's St. John jacket, or Armani or whatever—and I thought, Dear God, this is the end. And she just went, 'You have to save my son, he's going to die!' And Bettilou said, 'I'll do whatever you want,' and started crying."

In February, 2001, Congress held the hearing, with Senators Arlen Specter and Paul Wellstone leading the panel. The gallery was filled with muscular-dystrophy patients and their families. An N.I.H. director testified about what the agency had been doing for Duchenne—which was, the senators learned, not very much. "I'm sure it was very unpleasant," Sweeney, who also spoke that day, says. "It looked very unpleasant from where I was sitting." Furlong, too, testified before the panel. Speaking in a strong voice, she told the senators that Duchenne "gets only one one-thousandth of the N.I.H. budget" and called for an investment of a hundred million dollars over the next five years. She also told a brief but potent anecdote: "One day long ago, my son Patrick was trying to convince me of a crazy argument he had. He said to me, 'Mom, pretend I'm in a midlife crisis.' In fact, he was. He was eight." The bill passed the Senate and the House that spring, and in December, 2001, President Bush signed the Muscular Dystrophy CARE Act into law.

"That changed the landscape entirely," Sweeney told me. With millions of dollars available for research, scientists began to concentrate their attention on Duchenne. In 2000, there were fewer than ten trials and studies on Duchenne in process; there are now more than fifty. Parent Project has a membership of three thousand families in the United States, with affiliates throughout Europe, and an annual budget of five million dollars. Because the organization still has relatively small amounts of money to spend—usually two hundred thousand to three hundred thousand dollars for any given treatment—Sweeney and Furlong often invest in small laboratories or drug companies that need startup capital to bring a drug to a point where it can attract larger investment.

In 2007, Parent Project began working with a small company in New Jersey, PTC Therapeutics, which was developing a drug called ataluren, intended for boys whose Duchenne is the result of a "nonsense mutation" in the gene—a period in the genetic code's sentence, which renders it unreadable by the body's protein-making machinery. Furlong first learned of the company, from Sweeney, in 2000, and offered to help fund development of the drug. But the mutation addressed by ataluren accounted for only about fifteen per cent of all Duchenne sufferers. Stuart Peltz, the C.E.O. of PTC Therapeutics, told me, "After a couple of years working together, Pat said to me, 'I love what you're doing for the fifteen per cent of the patients—now you've got to do something for the rest.'" Furlong said that she would raise a million dollars from Parent Project's membership to support the company's efforts to find other targets on the gene. Four promising new targets turned up, and the N.I.H. awarded PTC Therapeutics and Parent Project, along with the University of Pennsylvania, $15.4 million for further research. Peltz says that the partnership between a drug company, a patient-advocacy group, and government "set up a new paradigm in how to fund drugs for these rare diseases."

In April, 2008, ataluren went to blind clinical trial, with a hundred and seventy-four boys at clinics in North America, Europe, Australia, and Israel. Two groups of boys were given the drug, in a high and a low dose, and a third group was given a placebo. Over forty-eight weeks, their progress was monitored by six-minute walk tests, a standard measure of muscle function in Duchenne boys. To the disappointment of everyone involved, boys on the high dose of ataluren declined at the same rate as boys not on the drug. The clinical trial was suspended. But researchers noted that boys on the lower dose did not decline in function. This tallied with preclinical experiments on cell cultures, which showed that if the dose is too high the drug's effect shuts off. "We initially discounted this as some artifact of cell culture and not real," Sweeney says. Furlong has advocated for patients to be allowed to take the drug as part of an F.D.A.-approved "access program," which permits terminal patients to take an experimental drug when no other treatment is available.

Sweeney and Furlong are optimistic that it will benefit the boys. Furlong, who has seen more than a thousand Duchenne patients, and who, according to Sweeney, "understands the disease a lot better than a lot of the clinicians and scientists who work on it," is convinced that she noticed an effect in boys on ataluren. "I am well aware there could be a placebo effect," she says, "but I saw changes, a posture that changed, a gait that improved, energy." Sweeney, though, believes that a cure is decades away and will likely involve stem-cell therapy. "It's not just fixing one muscle," he told me. "It's fixing every muscle in the body. That's the problem. Getting the cells to the right place and then getting them to do the right thing—it's a daunting engineering problem as much as anything else." For now, he says, researchers are focussed on developing multiple therapies that address various pathways of the disease. "It's the same sort of idea as with AIDS, where you have a combination of drugs, where none of them cure you but they can hold the disease at bay quite successfully with many patients. That's the goal here."

* * *

ONE MORNING IN EARLY NOVEMBER, Furlong had a meeting at Cincinnati Children's Hospital, where her sons originally received their diagnosis. After that, Furlong had not returned to Cincinnati Children's for almost a decade, but in 1999 the hospital hired a new pediatric neurologist, Dr. Brenda Wong, and Furlong began to meet with her regularly. She urged Wong to build a department that would focus on multidisciplinary care and treatment for Duchenne boys—epidemiologists to care for hormonal problems that arise with steroid treatment, lung specialists, immunologists, cardiologists. Thanks in part to Furlong's efforts, which included fund-raising for the hospital, Cincinnati Children's is now the leading clinical treatment center for Duchenne in the country. "I'm happy that this happened here," Furlong told me, when I met with her at the hospital. "Because it's payback."

Furlong was sitting in the airy cafeteria with a pediatric cardiologist, Dr. Linda Cripe, an open-faced blond woman in her forties. They became acquainted almost ten years ago, when Furlong called Cripe to persuade her to attend Parent Project's annual meeting, in Pittsburgh. The day before the meeting, Cripe's flight was cancelled because of bad weather, but, having promised Furlong she would be there, she drove from Cincinnati to Pittsburgh overnight. "I'm cursing her the entire way," Cripe told me. "I'm so angry, saying, 'Who is this woman? She's making me go to this thing . . .'"

Doctors who attend Furlong's conferences often report a kind of conversion experience. Jonathan Finder, the pulmonologist, addressed the Parent Project annual meeting in 1999. He told me, "It was one of these things that are hard to describe, the emotional impact of being surrounded by parents. As a physician, you meet with parents one at a time; you don't go into rooms filled with them." Cripe, at her first conference, spoke to a crowd of three hundred and fifty parents. Afterward, she called the hospital's other senior cardiologist and persuaded him to focus more on Duchenne patients. "Pat forced me to go into a situation that I didn't want to be in," Cripe says. "She brought the problem to me. I think if you're paying attention in medicine you'd have to be deaf, dumb, and blind not to see that there was a message there."

Three years ago, Cripe was instrumental in bringing about one of the first pediatric heart transplants given to a Duchenne child. During my visit, the recipient of the transplant, Riley Herrera, from Helena, Montana, was at Cincinnati Children's for a follow-up, and Furlong met him and his father, Ron, for lunch. Ron, a federal law-enforcement agent, is a trim, muscular man with a military bearing. Riley has dark hair, braces on his teeth, and a shy smile. He has been on steroids since the age of seven, and at eighteen is smooth-faced and under five feet tall; he might have been mistaken for a twelve-year-old. He was still walking, albeit with the characteristic Duchenne gait; we went to lunch at a nearby Mexican restaurant, and his father had to lift him up the two stairs that led inside.

At lunch, Ron asked about the latest developments in treatment. "The myostatin—is that a good one?" he said. He was referring to a drug, being developed by a company called Acceleron Pharma, that promotes muscle growth by inhibiting the uptake of myostatin, a protein that the body produces to prevent muscles from growing too large.

"Well, we'll see what Acceleron does, right?" Furlong said, carefully. With parents, she dispenses what she calls "therapeutic doses of hope" but does not promise too much. The drug is still in the early stages of testing and far from approval by the F.D.A.

"I saw a video of that dog that had the exon-skipping," Ron said. A recent YouTube clip showed a puppy that had been treated with a drug for a Duchenne variety caused by gaps in the code of the dystrophin gene. The drug had shown some success in dogs, but it would be years before it could be used in humans.

"Oh, yeah," Furlong said, in a neutral voice.

Talk turned to Riley's heart transplant. In 2006, during a family trip to New York, Riley had turned pale and become too tired to walk. "I'd get stabbing pains in my stomach," Riley recalled. "It was hard to breathe." When the family got home, Ron took him to a cardiologist in Billings, who said that Riley's heart was failing and that only a transplant could save him. Ron, who had been a member of Parent Project since 1999, turned to Furlong. She approached Cincinnati Children's, but there was controversy about giving a transplant to a "terminal" child, since pediatric hearts were in such short supply. Cripe argued that Riley should be on the transplant list. "The definition of Duchenne includes loss of ambulation between eight and ten years," Furlong explained. "Riley was fourteen and walking, so by clinical definition he was not a typical Duchenne." The hospital agreed to put Riley on the list, and Furlong found a donor to pay for him and his family to live in Cincinnati while they waited for a pediatric heart to become available.

Riley arrived in Cincinnati in January, 2007. Three months later, no heart had arrived. Furlong, who visited Riley in the hospital one Friday at the end of March, knew that he was running out of time. His heart was so enlarged that she could see it beating beneath his rib cage, through his shirt. This time, there was nothing that she could do.

"You said, 'You're going to need a heart in the next day or two,'" Ron said to Furlong. "And it happened—I think the next day!"

"The next day," Furlong said. "It was Friday. The heart came Saturday."

Riley was looking closely at Furlong during this exchange, apparently learning for the first time how desperate his situation had been. Had it not been for Cripe—and for Furlong's long campaign to reform Cincinnati Children's Hospital—he would almost certainly have died. He said, "So—you're like an angel."

Furlong scoffed. "Oh, yeah," she said, chuckling. "Tell my husband."

After lunch, Furlong was due to drive home to Middletown. Despite years of difficulty, she and Tom have repaired their relationship. "He knows why I do what I do and is happy for progress," Furlong says, "but it is still an open wound. I think he knew that this family needed both of us, even if we hated each other for a time." Before Furlong left, I asked her about Parent Project's long-term goals. "It's about looking at the landscape and seeing where you can play in this chess match, see how you can maximize the impact on research, see how you can apply that directly and quickly to the kids," she said. "From the day my sons were diagnosed, I wanted to buy time. So we look at a research opportunity and say, 'What is going to buy us time? What do you think we could tweak, invest in, move, that would buy us more time?' "

CHARLES HOMANS

Hot Air

FROM *COLUMBIA JOURNALISM REVIEW*

A significant number of TV weathercasters reject global warming, some going so far as to label it a scam. Charles Homans investigates why these close observers of weather are so skeptical.

THE SMALL MAKEUP ROOM OFF THE MAIN FLOOR OF KUSI's studios, in a suburban canyon on the north end of San Diego, has seen better days. The carpet is stained; the couch sags. John Coleman, KUSI's weatherman, pulls off the brown sweatshirt he has been wearing over his shirt and tie all day and appraises himself in the mirror, smoothing back his white hair and opening a makeup kit. "I kid that I have to use a trowel, to fill the crevasses of age," he says, swiping powder under one eye and then the other. "People have tried to convince me to use more advanced makeup, but I don't. I don't try to fool anyone."

Coleman is seventy-five years old, and looks it, which is refreshing in the Dorian Gray–like environs of television news. He refers to his position at KUSI, a modestly eccentric independent station in San Diego whose evening newscast usually runs fifth out of five in the local market, as his retirement job. When he steps in front of the green screen, it's clear why he has chosen it over actual retirement; in front of the camera he moves, if not quite like a man half his age, then at least like a man three quarters of it. His eyes light up, and the slight stoop with which he otherwise carries himself disappears. His rumble of a voice evens out into a theatrical baritone, full of the practiced jocularity of someone who has spent all but the first nineteen years of his life on TV.

By his own rough estimate, John Coleman has performed more than a quarter million weathercasts. It is not a stretch to say that he is largely responsible for the shape of the modern weather report. As the first weatherman on ABC's *Good Morning America* in the late 1970s and early '80s, Coleman pioneered the use of the on-screen satellite technology and computer graphics that are now standard nearly everywhere. In 1982, chafing at the limitations of his daily slot on GMA, Coleman used his spare time—and media mogul Frank Batten's money—to launch The Weather Channel. The idea seemed quixotic then, and his tenure as president ended a year later after an acrimonious split with Batten. But time proved Coleman to be something of a genius—the channel was turning a profit within four years, and by the time NBC-Universal bought it in 2008 it had 85 million viewers and a $3.5 billion price tag.

Those were the first two acts of Coleman's career. On a Sunday night in early November 2007, Coleman sat down at his home computer and started to write the 967 words that would launch the third. "It is the greatest scam in history," he began. "I am amazed, appalled and highly offended by it. Global Warming: It is a SCAM."

What had set him off was a football game. The Eagles were playing the Cowboys in Philadelphia on *Sunday Night Football*, and as a gesture of environmental awareness—it was "Green is Universal" week at NBC-Universal—the studio lights were cut for portions of the pre-game and half-time shows. Coleman, who had been growing increasingly skeptical about global warming for more than a decade, finally snapped. "I couldn't take it anymore,"·he told me. "I did a Howard Beale."

Skepticism is, of course, the core value of scientific inquiry. But the essay that Coleman published that week, on the Web site icecap, would have more properly been termed rejectionism. Coleman wasn't arguing against the integrity of a particular conclusion based on careful original research—something that would have consti-tuted useful scientific skepticism. Instead, he went after the motives of the scientists themselves. Climate researchers, he wrote, "look askance at the rest of us, certain of their superiority. They respect government and disrespect business, particularly big business. They are environmentalists above all else."

The Drudge Report picked up Coleman's essay, and within days its author was a cause célèbre on right-wing talk radio and cable tele-vision, beaming into Glenn Beck's TV show via satellite from the KUSI studios to elaborate on the scientists' conspiracy. "They all have an agenda,"·Coleman told Beck, "an environmental and politi-cal agenda that said, 'Let's pile on here, we're all going to make a lot of money, we're going to get research grants, we're going to get awards, we're going to become famous.' "

· Along with the appearances on Beck's and Rush Limbaugh's pro-grams came speaking offers, and soon Coleman was on the confer-ence circuit, a newly minted member of the loose-knit confederation of professional skeptics. (Coleman insists his views on climate change are apolitical, and says he has turned down offers to speak at Tea Parties and other conservative events.) His interviews and speeches that have been posted to YouTube have, in some cases, been viewed hundreds of thousands of times.

None of it would have had much of an impact, but for Coleman's résumé. For the many Americans who don't understand the difference between weather—the short-term behavior of the atmosphere—and climate—the broader system in which weather happens—Coleman's professional background made him a genuine authority on global warming. It was an impression that Coleman encouraged. Global warming "is not something you 'believe in,' " he wrote in his essay. "It is science; the science of meteorology. This is my field of life-long expertise."

Except that it wasn't. Coleman had spent half a century in the trenches of TV weathercasting; he had once been an accredited meteorologist, and remained a virtuoso forecaster. But his work was more a highly technical art than a science. His degree, received fifty years earlier at the University of Illinois, was in journalism. And then there was the fact that the research that Coleman was rejecting wasn't "the science of meteorology" at all—it was the science of climatology, a field in which Coleman had spent no time whatsoever.

COLEMAN'S CRUSADE CAUGHT THE EYE of Kris Wilson, an Emory University journalism lecturer and a former TV news director and weatherman himself, and Wilson got to wondering. He surveyed a group of TV meteorologists, asking them to respond to Coleman's claim that global warming was "the greatest scam in history." The responses stunned him. Twenty-nine percent of the 121 meteorologists who replied agreed with Coleman—not that global warming was unproven, or unlikely, but that it was a scam. Just 24 percent of them believed that humans were responsible for most of the change in climate over the past half century—half were sure this wasn't true, and another quarter were "neutral" on the issue. "I think it scares and disturbs a lot of people in the science community," Wilson told me recently. This was the most important scientific question of the twenty-first century thus far, and a matter on which more than eight out of ten climate researchers were thoroughly convinced. And three quarters of the TV meteorologists Wilson surveyed believe the climatologists were wrong.

In fact, anecdotal evidence of this disconnect had been accruing for several years. When a freakish snowstorm hit Las Vegas in December 2008, CNN meteorologist Chad Myers, appearing on *Lou Dobbs Tonight,* used the occasion to expound on his own doubts about global warming. "You know, to think that we could affect weather all that much is pretty arrogant," he told Dobbs. "Mother Nature is so big, the world is so big, the oceans are so big." Today's most oft-quoted and influential skeptics include Joseph D'Aleo, The Weather Channel's first director of meteorology, and Anthony Watts, a former Chico, California, TV meteorologist and prolific blogger who is leading a volunteer effort to document irregularities among the twelve hundred weather stations the National Weather Service maintains across the country (a concern that the National Oceanic and Atmospheric Administration considers negligible, and in any case has factored into its calculations since the '90s). When Oklahoma Senator James Inhofe, Congress's most reliable opponent of climate-change legislation, presented a list of more than four hundred "science authorities" who disagreed with the prevailing scientific opinion on climate change in 2008, forty-four of them were TV weathercasters. And after the signature of Mike Fairbourne, the weatherman for Minneapolis's CBS affiliate, turned up on a similar petition that year, reporters for the Minneapolis *Star Tribune* called around and found that hardly any of the city's TV weathercasters believed in climate change; one had recently called the idea "crazy" on a local talk-radio show.

More striking is the fact that the weathercasters became outspoken in their rejection of climate science right around the time the rest of the media began to abandon the on-the-one-hand, on-the-other-hand approach that had dominated their coverage of the issue for years, and started to acknowledge that the preponderance of evidence lay with those who believed climate change was both real and man-made. If anything, that shift radicalized the weathermen. "I think the media is almost sleeping with the enemy," one meteorologist told me. "The way it is now, there is just such a bias as to what gets out."

Free-market think tanks like the Heartland Institute, knowing an opportunity when they see one, now woo weathercasters with invitations to skeptics' conferences. The National Science Foundation and the Congress-funded National Environmental Education Foundation, meanwhile, are pouring money into efforts to figure out where exactly the climate scientists lost the meteorologists, and how to win them back. The American Meteorological Society (AMS)— which formally endorsed the scientific consensus on climate change years ago, but counts many of the skeptics among its members, to its chagrin—has started including climate-change workshops for weathercasters in its conferences.

For all of their differing agendas, the outfits have one thing in common: they have all realized that, however improbably, the future of climate-change policy in the United States rests to a not-insubstantial degree on the well-tailored shoulders of the local weatherman.

IN THE FALL OF 2008, researchers from George Mason and Yale universities conducted the most fine-grained survey to date about what Americans know and think about climate change. The short answer, unsurprisingly, was not very much. "Climate change is an incredibly complicated subject," says Anthony Leiserowitz, director of the Yale Project on Climate Change and one of the study's co-authors. "Most people are not interested in digging through the scientific literature, and in that situation trust becomes an enormous factor. We rely on people and organizations to guide us through this incredibly complicated and risky landscape."

That was where the survey's findings got interesting. When asked whom they trusted for information about global warming, 66 percent of the respondents named television weather reporters. That was well above what the media as a whole got, and higher than the percentage who trusted Vice-President-turned-climate-activist Al Gore, either of the 2008 presidential nominees, religious leaders, or corporations. Scientists commanded greater credibility, but only 18 percent of Americans actually know one personally; 99 percent, by contrast, own a television. "Meteorology benefits from the fact that we're just about the only science that has an individual in people's living rooms every night," says Keith Seitter, the executive director of the American Meteorological Society. "For many people, it's the only scientist whose name they know."

There is one little problem with this: most weathercasters are not really scientists. When Wilson surveyed a broader pool of weathercasters in an earlier study, barely half of them had a college degree in meteorology or another atmospheric science. Only 17 percent had received a graduate degree, effectively a prerequisite for an academic researcher in any scientific field.

This case of mistaken identity has been a source of tension throughout television's sixty-odd-year history. When TVs began to proliferate in postwar American households, the first generation of weathercasters that viewers saw on them was mostly military men, recently discharged World War II veterans who had trained in meteorology in the Navy and the Army Air Corps. (Louis Allen, Washington, D.C.'s first TV weatherman, had drawn up the forecasts for the invasions of Iwo Jima and Okinawa.) But as broadcasting licenses multiplied and stations began to compete with each other in the '50s, meteorologist Robert Henson recounts in *Weather on the Air: A History of Broadcast Meteorology* (to be published this year), the Army men gave way to entertainers: scantily clad "weather girls" abounded, as did puppets, including one who divined the forecast with his handlebar mustache. A weatherman in Nashville read his forecast in verse. One New York station featured a "weather lion."

After a few years of this sort of thing, the American Meteorological Society decided to step in; the professional association's membership, then comprised mostly of government and academic meteorologists, had grown wary of what the weather girls were doing to their reputation. The society devised a voluntary meteorological certification system, a seal of approval that TV weathercasters could obtain with the right academic background—at least a bachelor's degree in meteorology—or demonstrated knowledge in the field. (This seal is what technically distinguishes a meteorologist from a weathercaster.) In a 1955 *TV Guide* article entitled "Weather Is No Laughing Matter," AMS member Francis Davis wrote that "If TV weathermen are going to pose as experts, we feel they should be experts."

Although it took years, Davis's view eventually won out. By the end of the '70s, weathercasters had begun to treat their responsibilities with some seriousness. They started to see themselves as everyman (they were still mostly men) scientists, authority figures who helped viewers not only anticipate once-unpredictable events, but also comprehend them. And when you think about it, the achievement weathercasters have pulled off as science educators is remarkable— ask anyone with a television to name some meteorological terms, and odds are they will be able to rattle off half a dozen: low pressure systems, wind shear, cumulonimbus clouds. Weathercasters are usually a sort of science ambassador to their communities as well, and spend as much time talking to elementary school classes and civic groups about science as they do forecasting on the air. The work hasn't gone unappreciated; heaps of audience research have identified the weather report as the most popular segment of the local news broadcast, and the biggest factor in viewers' choice of which newscast to watch. Even as Americans' trust in the media as a whole has cratered, love for the weatherman has persisted at levels unchanged since Walter Cronkite's day.

The Clinton administration had all of this in mind in October

1997, when it gathered meteorologists from dozens of the nation's biggest television markets at the White House for a special summit on climate change. In two months, negotiators would be meeting in Kyoto to renegotiate the United Nations Framework Convention on Climate Change, the talks that would ultimately produce the Kyoto Protocol. Americans were still largely uninformed about climate change, and the White House was hoping the weathercasters could help bring them up to speed. More than one hundred of them showed up to hear speeches from Gore—an early version of the slideshow later documented in *An Inconvenient Truth*—and President Bill Clinton, as well as leading NOAA climate researchers.

As the administration had hoped, the meteorologists used the occasion to opine about climate change—but what many of them said wasn't quite what Al Gore had in mind. "There's still a significant segment of the scientific community that's not sold on this," Harvey Leonard, then the weatherman at WHDH in Boston, told *The Washington Post*. Others loudly refused to attend the summit, including all but one of the weathercasters in the Oklahoma City market. "I'm not smart enough to know [if the earth is warming], and I don't think any person on the planet is," KOKH meteorologist Tim Ross told the *Daily Oklahoman*. The following month, twenty TV weather personalities added their names to the Leipzig Declaration, a petition opposing the global warming theory.

It was only a blip on the radar, but it presaged the broader rejection of climate science that would come a decade later. The question was, why? No doubt, some of the blame belonged to the White House. In positioning themselves as advocates for not only a policy position but also a scientific one, Clinton and Gore had conflated the political question of what to do about climate change—one that was, and remains, deeply partisan in the U.S.—with the apolitical question of whether it was happening. This put the weathermen in a tricky spot—embracing what was, even then, the majority position in the scientific community would make them look like shills for the administration. "Since the White House is behind it, it's political," Leonard told the *Post*. "I'm not a lap dog," Gary England of KWTV in Oklahoma City—now a prominent climate skeptic—told the *Daily Oklahoman*. "I think Al Gore's motives were pretty good—he saw early on the potential that these people had," Kris Wilson says. "But he was probably the wrong spokesman. As journalists, we're taught to be skeptical, right? We're taught that if your mother says she loves you, get a second source."

But the disagreement, then as now, also came down to the weathercasters themselves, and what they knew—or believed they knew. Meteorology has a deceptively close relationship with climatology: both disciplines study the same general subject, the behavior of the atmosphere, but they ask very different questions about it. Meteorologists live in the short term, the day-to-day forecast. It's an incredibly hard thing to predict accurately, even with the best models and data; tiny discrepancies matter enormously, and can pile up quickly into giant errors. Given this level of uncertainty in their own work, meteorologists looking at long-range climate questions are predisposed to see a system doomed to terminal unpredictability. But in fact, the basic question of whether rising greenhouse gas emissions will lead to climate change hinges on mostly simple, and predictable, matters of physics. The short-term variations that throw the weathercasters' forecasts out of whack barely register at all.

This is the one explanation that everyone who has mulled the question seems to agree on—and indeed, when I spoke with meteorologists who were skeptical of or uncertain about the scientific consensus, it was the one thing they all brought up. "Meteorologists know our models," Brian Neudorff, a meteorologist at WROC in Rochester, New York, told me. "There's a lot of error and bias. We'll use five different models and come back with five different things. So when we hear that climatological models are saying this, how accurate are they?"

But that hardly explains why so many meteorologists have disregarded the mountain of evidence of global warming that has already occurred—or why, in the case of the hard-line skeptics, they are so fixated on proving a few data sets' worth of tree-ring and ice core measurements wrong. "I think a lot of people have theories," Robert Henson says, "but nobody knows for sure."

In the absence of a clear answer, several institutions— the National Environmental Education Foundation (NEEF), the Yale Forum on Climate Change & the Media, and the University Corporation for Atmospheric Research among them—have decided that education is the problem, and have launched projects aimed at teaching the weathercasters the basics of climatology. All proceed from the assumption that unreachable skeptics like Coleman are few and far between, and that most meteorologists are more uncertain than adamant, lost amid the Internet's slurry of fact and counter-fact. "While there is a group that seems to have made up their mind about climate change, there's still a substantial portion that's interested in learning more," says Sara Espinoza, a program director at NEEF. The AMS—which finds its credibility threatened by its televised emissaries a second time—is working with NEEF on a do-it-yourself climate science education package for meteorologists that points them to government data and peer-reviewed research. It is part of the AMS's broader "station scientist" program, which aims to give meteorologists the tools they need to become the go-to authorities in their newsrooms on all scientific subjects, not just the weather. In essence, it is a doubling down on the wager that the AMS made fifty-five years ago: if viewers are going to assume weathercasters are experts anyway, we might as well try to make them experts.

It remains a laudable goal. But in my own conversations with skeptical meteorologists, I began to think that that earlier effort had helped create the problem in the first place.

The AMS had succeeded in making many weathercasters into responsible authorities in their own wheelhouse, but somewhere along the way that narrow professional authority had been misconstrued

as a sort of all-purpose scientific legitimacy. It had bolstered meteorologists' sense of their expertise outside of their own discipline,
without necessarily improving the expertise itself. Most scientists
are loath to speak to subjects outside of their own field, and with
good reason—you wouldn't expect a dentist to know much about,
say, the geological strata of the Grand Canyon. But meteorologists,
by virtue of typically being the only people with any science background at their stations, are under the opposite pressure—to be
conversant in anything and everything scientific. This is a good
thing if you see yourself as a science communicator, someone who
sifts the good information from the bad—but it becomes a problem
when you start to see scientific authority springing from your own
haphazardly informed intuition, as many of the skeptic weathercasters do. Among the certified meteorologists Wilson surveyed in 2008,
79 percent considered it appropriate to educate their communities
about climate change. Few of them, however, had taken the steps
necessary to fully educate themselves about it. When asked which
source of information on climate change they most trusted, 22 percent named the AMS. But the next most popular answer, with 16
percent, was "no one." The third was "myself."

The biggest difference I noticed between the meteorologists who
rejected climate science and those who didn't was not how much
they knew about the subject, but how much they knew about how
much they knew—how clearly they recognized the limits of their
own training. Among those in the former category was Bob Breck,
the AMS-certified chief meteorologist at Fox affiliate WVUE in New
Orleans and a thirty-two-year veteran of the business. Breck rejected
the notion of human-driven climate change wholesale—"I just find
that [idea] to be quite arrogant," he told me. Instead, when Breck
talked to local schools and Rotaries and Kiwanis clubs about climate
change, he presented his own ideas: warming trends were far more
dependent on the water vapor in the atmosphere than carbon dioxide, he told them, and the appearance of an uptick in global tem-

peratures was the result of the declining number of weather stations in cold rural areas.

These theories were not only contradictory of each other, but had also been considered and rejected by climate researchers years ago. But Breck didn't read much climate research; "the technical journals are controlled by the professors who run the various societies," he told me, and those professors were hopelessly dependent on the "gravy train of grants from the NSF" that required them to propagate "alarmist theories." When I mentioned the AMS, Breck bristled. "I don't need the AMS seal—which I have," he said. "I don't need their endorsements. The only endorsements I need are my viewers, and they like what I do."

As Breck went on, I began to get a sense of the enormity of the challenge at hand. Convincing someone he is an expert is one thing. Actually making him one—well, that is another thing entirely.

CARL ZIMMER

The Singularity

FROM *PLAYBOY*

> *Brains that are enhanced by software. Or electrodes. Or genetic therapy. Or algae. Is this our future? Carl Zimmer takes a peek at what some visionaries call "the Singularity."*

L ET'S SAY YOU TRANSFER YOUR MIND INTO A
computer—not all at once but gradually, having electrodes inserted into your brain and then wirelessly out-sourcing your
faculties. Your vision is rerouted through cameras, your memories
are stored in a net of microprocessors and so on, until at last the
transfer is complete. As neuroengineers get to work boosting the
performance of your uploaded brain so you can now think as a god,
your fleshy brain is heaved into a bag of medical waste. As you—for
now let's just call it "you"—start a new chapter of existence exclu-

sively within a machine, an existence that will last as long as there are server farms and hard-disk space and the solar power to run them, are "you" still actually you?

This question was being considered carefully and thoroughly by a 43-year-old man standing on a giant stage backed by high black curtains. He had the bedraggled hair and beard of a Reagan-era metalhead. He wore a black leather coat and an orange-and-red T-shirt covered in stretched-out figures from a Stone Age cave painting.

He was not, in fact, insane.

The man was David Chalmers, one of the world's leading philosophers of the mind. He has written some of the most influential papers on the nature of consciousness. He is director of the Centre for Consciousness at Australian National University and is also a visiting professor at New York University. In other words, he has his wits about him.

Chalmers was speaking midway through a conference in New York called Singularity Summit 2009, where computer scientists, neuroscientists and other researchers were offering their visions of the future of intelligence. Some ideas were tentative, while others careened into what seemed like science fiction. At their most extreme the speakers foresaw a time when we would understand the human brain in its fine details, be able to build machines not just with artificial intelligence but with super intelligence and be able to merge our own minds with those machines.

"This raises all kinds of questions for a philosopher," Chalmers said. "Question one: Will an uploaded system be conscious? Uploading is going to suck if, once you upload yourself, you're a zombie."

Chalmers didn't see why an uploaded brain couldn't be conscious. "There's no difference in principle between neurons and silicon," he said. But that led him to question number two: "Will an uploaded system be me? It's not a whole lot better to be conscious as someone else entirely. Good for them, not so good for me."

To try to answer that question Chalmers asked what it takes to be me. It doesn't take a certain set of atoms, since our neurons break down their molecules and rebuild them every day. Chalmers pondered the best way to guarantee the survival of your identity: "Gradual uploading is the way to go, neuron by neuron, staying conscious throughout."

But perhaps that won't be an option. Perhaps you will have died by the time you are uploaded. Chalmers didn't find this as alarming as others do. "Let's call it the Buddhist view," he said. Every day, he pointed out, we lose consciousness as we fall asleep and then regain it the next morning. "Each waking is really like a new dawn that's a bit like the commencement of a new person. But it turns out that's good enough. That's what ordinary survival is. We've lived there a long time. And if that's so, then reconstructive uploading will also be good enough."

IF THE TERM SINGULARITY RINGS a bell, maybe you've read the 2005 bestseller *The Singularity Is Near.* Its author, computer scientist and inventor Ray Kurzweil, confidently predicts intelligence will soon cross a profound threshold. The human brain will be dramatically enhanced with engineering. Artificial intelligence will take on a life of its own. If all goes well, Kurzweil predicts, we will ultimately fuse our minds with this machine superintelligence and find a cybernetic immortality. What's more, the Singularity is coming soon. Many of us alive today will be part of that transformation.

The Singularity is not only a future milestone but also a peculiar movement today. Along with spaceflight tycoon Peter Diamandis, Kurzweil has launched Singularity University, which brought in its first batch of students in the summer of 2009. Kurzweil is also director of the Singularity Institute for Artificial Intelligence, which held its first annual summit in 2006. The summits are a mix of talks by Kurzweil and other Singularity advocates, along with scientists working on everything from robot cars to gene therapy. For its first three years the Singularity Summit took place around the Bay Area, but in 2009 the institute decided to decamp from its utopian environs and head for the more cynical streets of New York.

I was one of the curious skeptics who heeded the call and came to the 92nd Street Y. I've been writing about new advances in science for 20 years, and along the way I've developed a strong immune defense against hype. The Singularity, with all its promises of a techno-rapture, seems tailor-made to bring out the worst in people like me. The writer John Horgan, who has even less patience for the promises science cannot keep, wrote "Science Cult," a devastating essay about the Singularity, for Newsweek.com in May 2009.

He acknowledges part of him enjoys pondering the Singularity's visions, such as boosting your IQ to 1,000. "But another part of me—the grown-up, responsible part—worries that so many people, smart people, are taking Kurzweil's sci-fi fantasies seriously," he wrote. "The last thing humanity needs right now is an apocalyptic cult masquerading as science."

I decided to check out the Singularity for myself. The summit turned out to be one of the most bizarre experiences I've had. Chalmers wasn't the only speaker to induce hallucinations. Between the talks, as I mingled among people wearing S lapel pins and eagerly discussing their personal theories of consciousness, I found myself tempted to reject the whole smorgasbord as half-baked science fiction. But in the end I didn't.

After the meeting I visited researchers working on the type of technology that people such as Kurzweil consider the steppingstones to the Singularity. Not one of them takes his extreme vision of the future seriously. We will not have some sort of cybernetic immortality in the next few decades. The human brain is far too mysterious and computers far too crude for such a union anytime soon, if ever. In fact some scientists regard all this talk of the Singularity as nothing more than recklessly offering false hope to people currently struggling with blindness, paralysis and other disorders.

But when I asked these skeptics about the future, even their most conservative visions were unsettling: a future in which people boost their brains with enhancing drugs, for example, or have sophisticated computers implanted in their skulls for life. While we may never be able to upload our minds into a computer, we may still be able to build computers based on the layout of the human brain. I can report I have not drunk the Singularity Kool-Aid, but I have taken a sip.

THE FUTURE IS NOT NEW. By the dawn of the 20th century science was moving so fast many people were sure we were on the verge of tremendous change. The blogger Matt Novak collects entertainingly bad predictions at his website Paleo-Future. My favorite is a 1900 article by John Watkins that appeared in *Ladies' Home Journal*, offering readers a long list of predictions from leading thinkers about what life would be like within the next 100 years.

"A man or woman unable to walk 10 miles at a stretch will be regarded as a weakling," Watkins wrote. "There will be no C, X or Q in our everyday alphabet."

As science advanced through the 20th century, the future morphed accordingly. When scientists figured out how to culture animal cells in the early 1900s, some claimed such cells would let us live forever. In the 1940s the availability of antibiotics led some doctors to declare the age of infectious diseases over. The engineers who founded NASA were sure we would build cities on the moon, perhaps even Mars. And as scientists began to develop computers and the programs to run them, they began to predict that someday— someday soon—computers would gain a human intelligence.

Their confidence grew not just from their own research. Neuroscientists were discovering that our brains act a lot like computers, so it seems logical we would someday be able to translate the neural code, build computers that process information similarly and even join brains and machines together.

In 1993 the science-fiction writer Vernor Vinge wrote an essay on this particular kind of future. He entitled it "The Coming Technological Singularity," borrowing a term astrophysicists use to describe a place where the ordinary rules of gravity and other forces break down. "Within 30 years we will have the technological means to create superhuman intelligence," he wrote. "Shortly after, the human era will be ended."

By the late 1990s Kurzweil emerged as the leading champion of the coming end of life as we know it. He started as a tremendously successful computer scientist, having invented print-to-speech machines for the blind, music synthesizers and a host of other devices, and in 1990 he published his first forward-looking book, *The Age of Intelligent Machines*. He argues that within a few decades computers will be as intelligent as humans, if not more so.

As the years passed, his predictions grew more extreme. In his 1999 book, *The Age of Spiritual Machines*, he imagines life in 2099. "The number of software-based humans vastly exceeds those still using native neuron-cell-based computation," he wrote. In 2005 he brought Vinge's term to wide attention in his book *The Singularity Is Near*, in which he bemoans how hobbled we are by feeble neurons, bones and muscles. "The Singularity," he writes, "will allow us to transcend these limitations of our biological brains and bodies."

At Singularity Summit Kurzweil came onstage to offer the latest iteration of his case for the Singularity. The audience broke into fierce applause when he appeared, a few people standing and pounding their hands in slow motion. Kurzweil was, as ever, sharply dressed, wearing a tailored blue suit, an open striped shirt and narrow glasses. (In 1984 a writer for *Business Week* noted he "wears expensive Italian suits and a gold Mickey Mouse watch.")

He launched into his talk, leaning back on one foot as he spoke, his small frame angled diagonally to the audience, his eyebrows softly raised, his entire body seemingly caught in a perpetual shrug. Rather than slamming the audience with an infomercial pitch, his languid body language seemed to be saying, "Look, I don't care if you believe me or not, but these are the facts."

He talked about everything from quantum physics to comas to speech recognition, but at the heart of his talk was a series of graphs. They showed an exponential growth in the power of technology, from the speed of DNA sequencing efforts to the power of computers to the growth of the Internet. This exponential growth has been so relentless that Kurzweil has dubbed it the law of accelerating returns.

"It really belies the common wisdom that you can't predict the future," he said as he gazed at the graphs. Thanks to the law of accelerating returns, he said, technology will continue to leap forward and astonish us. "Thirty linear steps take you to 30," he said. "Thirty exponential steps take you to a billion."

ONE WAY TO JUDGE WHETHER Kurzweil is right about the future is to see how well he has done in the past. In the realm of computer science he has done pretty well. In 1990 he predicted the world chess champion would be a computer by 1998. (IBM's Deep Blue computer beat Garry Kasparov in 1997.) In 1999 Kurzweil predicted that in 10 years computers communicating with one another and the world wide web wirelessly would be commonplace. It's already becoming hard to recall when computers were lashed to modems.

But many of Kurzweil's predictions have failed. In 1999 he predicted that by 2009 "bioengineered treatments for cancer and heart disease will have greatly reduced the mortality from these diseases." In 2006, the most recent year for which statistics are available, 829,072 people died in the United States of heart disease. Fortunately the death rate from heart disease is lower now than in 1950, but that drop is due mainly to low-tech measures such as getting people to stop smoking. Meanwhile the death rate from cancer has dropped only five percent since 1950.

These failed predictions reveal a weakness at the heart of Kurzweil's forecasts. Scientific understanding doesn't advance in lockstep with technological horsepower. It was funny, in a morbid way, to watch Kurzweil make his case inside the 92nd Street Y just as a surge of swine flu viruses was sweeping the city . There was a time when sequencing the entire genome of a single flu virus was a colossal, budget-busting project; now it costs a few hundred dollars. As of October 2009 the Influenza Genome Sequencing Project has collected the complete genomes of 4,115 viruses from around the world, and that number is rising rapidly. All that raw genetic information has certainly allowed scientists to learn important things about the flu, but it has not given New York any fancy new way to stop it in its tracks. New Yorkers could only wash their hands as they waited for the delivery of new vaccines and hoped their hospitals didn't get overwhelmed by people in need of respirators. All thanks to a virus with just 10 genes.

Even as flu viruses multiplied through the city, Kurzweil happily continued his talk, mocking the skeptics who scoffed when scientists were trying to sequence all 3.5 billion "letters" of DNA in the human genome. For a long time it seemed as if they'd never finish, and then in the early 2000s they were done. Years later we still have lots of open questions about how the genome actually works. Scientists used to think it contained 100,000 protein-coding genes, but it turns out to have just 20,000, and researchers don't actually know the function of many of them. What's more, the genome also has tens of thousands of genes our cells use to make single-strand versions of DNA called RNA. Many play vital roles in the cell, but a lot probably don't do anything, and scientists are just starting to figure out what does what. Sequencing the human genome was promised by some to yield cures to most diseases. Scientists are now searching the genome for genes that increase your risk of high blood pressure, diabetes and other diseases. In most cases they have found lots of genes that raise your risk by only a barely measurable amount. Sequencing the human genome has revealed to scientists that they know less than they thought they did.

* * *

WHEN I STARTED CONTACTING EXPERTS about the Singularity, something surprising happened—they didn't laugh and hang up the phone.

"I find some people way too quick to pooh-pooh the idea of an impending Singularity," said Martha Farah, director of the Center for Cognitive Neuroscience at the University of Pennsylvania. Farah investigates how people try to enhance their cognition with drugs. Drugs originally designed to treat mental disorders are now being taken by perfectly healthy people. Adderall, a drug for ADHD, is a popular campus drug for boosting concentration. Modafinil, developed for people with narcolepsy, is now a drug of choice for those who want to burn the midnight oil.

In the years to come Farah anticipates even more powerful drugs will come to market. Some are intended to slow the disappearance of memories in people with Alzheimer's disease; others may boost cognition in people with impairments. She expects there will be people—maybe a lot of them—who will take these drugs in the hopes of boosting an already healthy brain, not to fix a deficit.

In December 2008 Farah and a group of fellow neuroscientists and bioethicists wrote in *Nature* that this kind of brain boosting is okay. "Mentally competent adults should be able to engage in cognitive enhancement using drugs," they wrote. It's impractical, they argued, to try to draw a line between treating a disease and enhancing a healthy brain.

Farah sees an urgent need now to measure the actual enhancement these drugs can bring. "The effectiveness of Adderall depends crucially on the individual," she said. "The literature suggests that people who are average or below get the biggest benefit. The high performers may get no benefit or may actually be impaired by it."

She is now measuring the performance of students on Adderall and placebos to see if that's actually the case.

Farah feels the drug dosing of today will have a profound impact on how we treat our brains: "I think this growing practice may be softening us up to accept more drastic brain modifications down the line."

"HERE ARE SOME CRITTERS," SAID Ed Boyden. The serene young leader of the Synthetic Neurobiology Group at MIT stood in his laboratory. Tiny pieces of electronics were strewn across the lab benches. Dishes full of neurons were positioned under microscopes.

And there were also flasks of algae, one of which Boyden had grabbed and held up for me to see. He sloshed the green fluid around. I came to Boyden's lab after seeing him give a remarkable talk at the Singularity Summit. Boyden is at the forefront of a field known as neuroengineering. Neuroengineers seek to restore damaged brains by implanting tiny on-board computers and electrodes. Boyden's research may take neuroengineering to a new level. Rather than use electricity to manipulate the brain, he wants to use light. But in order to do that, he will have to borrow some genes from the algae he was holding and put them in human DNA . If he succeeds, people will be part machine, but they will also be a little bit algae, too.

The logic behind brain implants is simple. Many disorders, from blindness to paralysis, come down to a break in the flow of signals through our nervous system. Neuroengineers have long dreamed of building tiny machines that could restore that flow. So far, they've had one great success: the cochlear implant, a machine that delivers sound to the brains of the deaf. A cochlear implant picks up sounds with an external microphone and converts them to electronic signals, which travel down wires into a shell in the ear called the cochlea, where they tickle the ends of the auditory nerves. The first generation of cochlear implants, in the 1970s, were big, awkward devices with wires crossing the skull, raising the risk of infection. They used up power quickly and produced crude perceptions of sound. In the 1990s scientists developed microphones small enough to perch on the ear that transmit sounds wirelessly to an implanted receiver. Today more than 180,000 people use cochlear implants. Scientists continue to make improvements to the implants so they can run on far less energy yet perform even better.

Neuroengineers have also been testing implants that go into the brain itself, but progress has been slower on that front. So far 30,000 people have had electrodes implanted in their brains to help them cope with Parkinson's disease. Pulses of electricity from the implants make it easier for them to issue the commands to move their bodies. Other scientists are experimenting with similar implants to treat other disorders. In October 2009 scientists reported 15 people with Tourette's syndrome had 52 percent fewer tics thanks to deep-brain stimulation. Other scientists are trying to build the visual equivalent of a cochlear implant. They've linked cameras to electrodes implanted in the visual centers of blind people's brains. Stimulating those electrodes allows the subjects to see a few spots of light.

While these electrodes send electricity into the brain, other units that researchers are working on pull information out. At Massachusetts General Hospital doctors have started clinical trials on human volunteers to test brain implants that give paralyzed people the ability to control a computer cursor with thought alone. Other neuroengineers have been able to achieve even more spectacular results on monkeys. At the University of Pittsburgh, for example, monkeys can use their thoughts to feed themselves with a robotic arm.

These are all promising results, but brain implants are still fairly crude. When the electrodes release pulses of electricity they can't target particular neurons; they just blast whatever neurons happen to be nearby. The best electrodes for recording brain activity, meanwhile, can pick up only a tiny portion of the chatter in the brain because engineers can implant only a few dozen electrodes in a single person.

Making matters worse, almost all of these implants have to be rigged to wires that snake out of the skull and draw a lot of power, relatively speaking, limiting battery lifetime. Surgeons have also found that the brain attacks these electrodes, covering them with a protective coat of cells that can render them useless. All of these problems mean you can't expect to carry a brain implant for life. Dipping into a person's brain from time to time to swap out implants and batteries would not just be expensive but would also pose the risk of infection.

But none of these challenges is necessarily a showstopper. Scientists are working on new designs that can allow brain implants to shrink in size, use less power, and deliver better performance. In 2009, for example, a team of scientists at MIT reported how they had implanted a tiny electrode into the brain of zebra finches. The birds could fly freely around an enclosure. But with the press of a button, the scientists could wirelessly transmit a signal to the song-producing region of the bird's brain. The bird instantly stopped singing.

Up to now, many brain implant studies have focused on delivering or receiving electrical signals. Boyden's approach is novel. Building implants that work with light and algae, it turns out, can make light-emitting devices possible. That's because some algae have channels on special membranes in their cells that respond to light of certain colors by opening up, allowing charged particles to move in or out of the membrane. A few years back Boyden wondered if he could insert those channels into neurons and use them as an optical switch. Hit a neuron with light and its channels will open, triggering a signal.

Boyden and his colleagues pinpointed the gene for the channel, inserted it into viruses and then put the engineered viruses into a dish of neurons. The viruses infected the neurons, and along with their own genes they inserted the light-channel gene from the algae. The virus is harmless, so the neurons did not suffer from the infection, but they started using the algae gene to build channels of their own. Boyden then exposed the neurons to a flickering blue light. The neurons responded by crackling with spikes of electricity. "We started playing around with it and we got light-driven spikes almost on the first try," he said. "It was an idea whose time had come."

Boyden and his colleagues published that experiment in 2005, and since then he has expanded his neuroengineer's tool kit dramatically. "We grow organisms to screen for new molecules," he said. Their discoveries of new light-sensitive channels in algae, bacteria and fungi have allowed them to engineer neurons that respond to a rainbow of light.

Boyden has neurons performing new tricks. He can get them to produce a voltage spike in response to light or to go completely quiet. He can flash a particular pattern of lights to trigger signals. He can also target his channels to different types of neurons by adding different genetic handles to the channel DNA. The genes get inserted into lots of cells, but they get switched on in only one kind of neuron. Flashing different colors of light, he can switch on and shut down different groups of neurons all at once.

Now he is starting to see how his engineered neurons behave in real brains rather than in petri dishes. One virus he selected for his experiments, known as an adeno-associated virus, has proven to be promising in human gene-therapy trials in other labs. It has safely delivered genes into the bodies of more than 600 people so far (though some of the genes have produced unintended negative side effects). Also, last April Boyden and his colleagues reported they were able to successfully infect certain neurons in the brains of monkeys without causing harm to the animals. The scientists inserted an optical fiber into the monkey brains and were able to switch the neurons on with flashes, just as they do in a petri dish.

Boyden gave a talk at the Singularity Summit, unveiling a particularly stunning experiment he and his colleague Alan Horsager have run at the University of Southern California on congenitally blind mice: They infected the animals with sight.

The mice were blind thanks to mutations in the light-receptor genes in their retinas. The team wondered if they could make those neurons sensitive to light again. They loaded genes for light-sensitive channels onto viruses and injected them into the mice. The genes were targeted for the retinal neurons missing their own light receptors. Boyden gave the mice enough time to incorporate the genes into their eyes and, he hoped, make the channels in their neurons. Since mice can't read eye charts out loud, he and his colleagues had to use a behavioral test to see if their eyes were working. They put the mice into a little pool with barriers arranged into a maze. At one end of the pool the mice could get out of the water by climbing onto an illuminated platform. Regular mice quickly followed the light to the platform while blind mice swam around randomly. The mice infected with neuron channels headed for the exit far more often than chance and almost as often as the healthy mice. Boyden and his colleagues have founded a company, Eos Neuroscience, to see if they can use this gene therapy to help restore some eyesight to people.

Ultimately, though, Boyden also wants to install these light-sensitive receptors on neurons deep in the brain. Then, with a flash of light, he can make certain neurons fire or go silent. Boyden and his colleagues have built a peculiar gadget that looks like a miniature glass pipe organ. At the base of each fiber a diode can produce an intense flash of light. Boyden envisions implanting this array into people's brains and then wirelessly programming it to produce a rapidfire rainbow pattern of light. His far-off goal is to help treat medical disorders with these implants, and he doesn't give much thought to the possibility of people using implants to enhance their brains, the way they do now with Adderall. Brain implants certainly inspire cool scenarios—what if someone wanted to see in ultraviolet or operate a jet fighter with thought alone?—but Boyden has the luxury of not having to worry about those ethical matters. After all, it's one thing to open a jar and pop a pill; it's quite another to undergo brain surgery. "I think the invasive techniques won't be used for augmentation for a long time to come," he says.

As the conversation continues we get to talking about Lasik. Once there was a time when having a laser shot into your eye to fix myopia was the ophthalmological equivalent of Russian roulette. "Forty years ago it was daring," says Boyden. "Now there are clinics that do hundreds of these day in and day out."

I OPENED MY SUMMIT SCHEDULE to see what was next.

9:35 A.M.: Technical Road Map for Whole Brain Emulation.

10:00 A.M.: The Time Is Now: We Need Whole Brain Emulation.

This should be interesting, I thought.

Beyond drugs and prosthetics is whole brain emulation. If you haven't heard of it, here's a quick definition from a 2008 paper by Nick Bostrom and Anders Sandberg, two scientists at the University of Oxford: "The basic idea is to take a particular brain, scan its structure in detail and construct a software model of it so faithful to the original that, when run on appropriate hardware, it will behave in essentially the same way as the original brain."

At the Singularity Summit, Sandberg strode onto the stage, wearing a science-fair smile and a bright red tie, to explain what it would take to reach that goal. First, scientists would have to decide exactly how much detail they'd need. Would they have to track every single molecule? Would a rough approximation of all 100 billion neurons suffice? Sandberg suspected scientists would need a scan of a brain that could provide details down to a few nanometers (less than a millionth of an inch). Today researchers at Texas A&M have figured out how to take images of the brain at a resolution of just 160 nanometers, but they've scanned only a rice-grain-size piece of mouse brain in any one trial. To scan the brain tissue the scientists must stain it with color-producing chemicals, dunk it in plastic for hardening and then shave away one layer at a time. For now brain emulation is a zero-sum game.

But let's assume for the moment scientists can scan an entire human brain at nanometer scale. It turns out the hard work is just beginning. Researchers will then have to write software that can turn all the data into a three-dimensional model and then boot up this virtual brain. Sandberg doesn't think a computer would have to calculate the activity of the neurons atom by atom. Neurons are fairly predictable, so it's already possible to build models that behave a lot like real neurons. Mikael Djurfeldt, a scientist at Sweden's Royal Institute of Technology, and his colleagues have succeeded in modeling one particular kind of neuron cluster known as a cortical column. They created a model of 22 million neurons joined together by 11 billion synapses. When their imitation neurons started to talk to each other, they behaved a lot like real cortical columns. Of course it's important to remember there are about 100 billion neurons in the human brain—several hundred times more than Djurfeldt has simulated. And even if scientists do manage to simulate 100 billion neurons, they'll also need to give the simulated brain a simulated world.

Sandberg has made some rough calculations of how much computing power that would demand and how fast computer power is rising, and he's fairly confident a whole-brain emulation will be possible in a matter of a few decades. Exactly how a whole-brain emulation would behave, he isn't sure.

"I don't know whether a complete simulation of a brain, one to one, would actually produce a mind," he said. "I find it pretty likely, but I don't have evidence for that. I want to test it."

Sandberg made an exit stage right, replaced at the podium by Randal Koene, a neuroscientist at the European technology firm Fatronik-Tecnalia. Koene offered some reasons for why anyone would want to work so hard to make a whole-brain emulation in the first place. Even if it just behaved like a generic human brain rather than my brain or yours in particular, scientists could still use it to run marvelous new kinds of experiments. They might test drugs for depression, Parkinson's and other disorders. Koene is also a strong advocate of so-called mind uploading—the possibility of not just running a brain-like simulation on a computer but actually transferring a person's mind into a machine. To him, it is the liberation of our species. "We must free the mind," said Koene.

For a little ground-truthing I called Olaf Sporns, a neuroscientist at Indiana University.

"This is not going to happen," he said.

Sporns is in a good position to judge. He and his colleagues have carried out just about the closest thing to whole-brain emulation given today's technology. They map human brains using a high-resolution method called diffusion spectrum imaging. They map the long fibers that link regions of the brain together like computers on the Internet. Sporns and his colleagues have analyzed the connections between 1,000 regions and have found the brain's network is organized according to some of the same rules of other large networks—including the Internet. For example, several regions act as hubs, while most regions are connected to only a few others. Sporns and his colleagues created a computer model of this brain network and let each region produce signals that could spread down the fibers. They found their simulation of a brain at rest produced distinctive waves that spread back and forth around the entire brain similar to the way waves spread across our own.

Whole-brain emulations will become more sophisticated in the future, said Sporns, but he finds it ridiculous to expect them to be able to capture an individual's mind. In fact, mind uploading is a distraction from the truly revolutionary impact whole-brain emulations will have. By experimenting with them, researchers may discover some of the laws for building thinking networks. It may turn out human brains can work only if their networks have certain arrangements. "It's like learning the laws of physics when you want to build an airplane. It helps," said Sporns.

Discovering those laws may allow computer scientists to finally build machines that have mental processes similar to ours. Sporns thinks scientists are already moving in that direction. IBM, for example, now has a contract with the military to "fabricate a multichip neural system of about 108 neurons and instantiate into a robotic platform performing at 'cat' level."

"That is going to happen," said Sporns.

This is the best I can manage for skeptics. Uploading your mind is science fiction. But endowing robots with a humanlike cognition? Only a matter of time.

"I HOPE YOU DIDN'T BELIEVE a word of that last talk."

I stood in a long line for the men's room during a coffee break. The women's room next door was empty. I thought how only at a meeting like the Singularity Summit would I find myself in this situation. When I heard someone talking to me, I turned to see a tousle-haired psychologist named Gary Marcus.

Good, I thought. Maybe Marcus would demolish the Singularity and leave nothing behind but smoking wreckage.

Marcus, who teaches at New York University, has spent years studying computation.

Computers are good at certain kinds of computations such as sorting things into categories. But they're not good at the things we do effortlessly, such as generating rules from experience.

Marcus was annoyed by one of the talks at the Summit, in which a computer scientist promised that humanlike artificial intelligence was nigh. "Figuring this stuff out in 10 years—I don't believe it," he said.

I expected some serious curmudgeonliness when Marcus delivered his talk the following day. In his 2008 book, *Kluge,* Marcus explored design flaws in the human brain. We don't simply store information on a hard disk, for example, but embed it in a web of associations. That's why memories may escape us until something— perhaps the taste of a cookie—brings back the right associations.

Marcus explains how these quirks are locked into our brains thanks to our evolutionary history. We did not evolve to be computers but animals that could learn to find food and avoid being eaten.

So I imagined Marcus would declare the human brain unimprovable, but he didn't. He stood up, explained the shortfalls of human memory and suggested memory would be a good place to start improving the human brain.

I called Marcus later and told him I was surprised. "Human enhancement is a real possibility," he replied. He thought a powerful way to enhance the brain would be with "cognitive prosthetics," a kind of onboard iPhone. The only challenge would be to decipher our brains' code well enough to let the iPhone talk to our brains. Marcus didn't see any reason scientists wouldn't eventually figure that out.

Like Sporns, Marcus agreed whole-brain emulations might turn out to be most valuable for what they reveal about the nature of intelligence. In the end that's what left Marcus worried. We just might succeed too well and program a machine with so much intelligence it can boost its own intelligence by itself.

"There are going to be machines that are cleverer than we are," he said.

It is time, Marcus says, to start planning for that world. It doesn't matter whether we live to see it or not; our grandchildren or great-grandchildren might. We owe it to them to get ready—if not for the Singularity then at least for a world different from our own.

BP's Deep Secrets

FROM *MOTHER JONES*

While the common perception of the 2010 Gulf oil spill may be that negative effects were minimal, the long-term environmental damage to the Gulf ecosystem may be catastrophic—as Julia Whitty finds out from scientists on the scene.

W E'RE SWINGING ON ANCHOR THIS AFTERNOON AS powerful bursts of wind blow down through the Makua Valley and out to sea. The gales stop and start every 15 minutes, as abruptly as if a giant on the flu side of the Hawaiian island of Oahu were switching a fan on and off. We sail at the gusts' mercy, listing hard to starboard, then snapping hard against the anchor chain before recoiling to port. The intermittent tempests make our work harder and colder. We shiver during the microbursts, sweat during the interludes, then shiver again from our own sweat.

I'm accompanying marine ecologist Kelly Benoit-Bird of Oregon State University, physical oceanographer Margaret McManus of the University of Hawaii-Manoa, and two research assistants aboard a 32-foot former sportfishing boat named *Alyce C*. On the tiny aft deck, where a marlin fisher might ordinarily strap into a fighting chair, Benoit-Bird and McManus are launching packages of instruments: echo sounders tuned to five frequencies; cameras; and a host of tools designed to measure temperature, salinity, current velocity, chlorophyll fluorescence, and zooplankton abundance, all feeding into computers lashed into the tiny forward cabin.

Despite the impressive technology crammed aboard the boat, its deployment is pure 19th century. At any given time, two of us man the aft winch, launching the equipment overboard by hand, feeding out dual lines of nylon and coaxial cable, slowly wearing calluses into our gloves as we ease the instruments through the water column at roughly 33 feet per minute. Six feet shy of the bottom, 74 feet down, the rig is hauled back up, collecting data the whole way. The process is repeated around the clock for the next 24 hours, a procedure either monotonous or meditative, depending on your frame of mind. Near the bottom, McManus calls, "Making a mark." She might as well be calling "mark twain."

But whereas old-time riverboat captains sounding with lead-weighted ropes were gleaning information about safe shipping channels and shifting sandbars, we're sounding for signs of life. To the untrained eye, the incoming echo soundings appear as waves of blue, green, and yellow scrolling horizontally across our computer monitors. To the trained eye, they appear as layers of life flooding in on darkness. Benoit-Bird points toward the screens, each one tuned to read the sonar signature of a different-size life form. "That layer is zooplankton," she says. "And that layer is fish." Suddenly, I can see a crude facsimile of the migrations of the nighttime sea.

Most of the marine life familiar to us at the surface inhabits the epipelagic zone, the sunlit realm, stretching down to about 600 feet. Yet many whales, dolphins, seals, sea turtles, sharks, manta rays, billfish, and smaller predatory fish are nocturnal hunters, dependent on the mysterious movements of a vast community of organisms known as the deep scattering layer, or DSL. This aggregation of life forms was unknown until the 1920s, when early hydrographers mapping the ocean with sound encountered a daytime "seafloor" around 3,300 feet, which rose perplexingly toward the surface at night. Named for its echo-reflecting signature, the DSL was eventually recognized by marine biologists in 1948 to be layers of living creatures hiding on the cusp between perpetual twilight and darkness.

What the echo sounders of old were actually picking up were the billions of swim bladders (buoyancy floats) of the fish inhabiting the dark realm of the DSL—primarily lantern fish, bristlemouths, and hatchetfish. These fish, generally between one and twelve inches long, are endowed with the usual fishy hardware of fins, scales, lateral lines, and tails. But their habit of hiding in the darkness by day and chasing darkness upward at night led to the development of extraordinarily large eyes and organs, known as photophores, capable of producing light—usually a weak blue, green, or yellowish light—the color and pattern of which signal the fish's species and gender, as well as information used in shoaling and other communications we don't understand. The photophores also create a camouflage known as counterillumination. By adjusting internal dimmer switches, these mesopelagic ("middle sea," or twilight zone) fish match the slightest overhead ambient light level—be it the faint glow of the sun or moon—making their silhouettes less visible to predators above and below.

DSL species rise at night—some to waters as shallow as 30 feet deep—for a variety of reasons: Some are avoiding the daytime surface hunters; others are avoiding the nocturnal hunters of the DSL who don't rise (like lancetfish); still others are saving energy by spending their days in a sleeplike state prompted by the frigid waters. (The alternative, living only at the warm surface, produces a fast metabolism requiring more food.) Krill, among the most abundant and important invertebrates of the DSL, rise at night to graze on the pastures of the sea: single-celled phytoplankton, plants that survive only in the sunlight zone.

The lantern fish, bristlemouths, hatchetfish, and crustaceans of the DSL are believed to account for 80 percent of all the biomass in the mesopelagic zone, with lantern fish alone making up some 660 million tons of living fish—perhaps the greatest distribution, population, and species diversity of all ocean fish on the planet. The mesopelagic fauna also includes many kinds of squid, krill, and siphonophores and ctenophores (jellyfish-like animals), as well as worms, sea butterflies, and larvae that comprise the DSL zooplankton. The vast life of the deep scattering layer supports the surface life above it, including the $172 billion global seafood and aquaculture industries.

It's no wonder then that most of the predators of the sunlit sea make their living diving to meet the DSL, which rises like a great dumbwaiter from the deep bearing every manner of seafood delicacy on a platter of darkness. No wonder, too, that the DSL is being eyed by the fishing industry as the last great resource to be exploited.

Not long after dark, dolphins show up on the data stream, monopolizing the monitors with bold red and orange signatures. These are spinner dolphins who've spent the daytime hours resting in shallow coastal waters, hiding from sharks, sleeping with eyes wide open and their echolocation shut down. During the couple of years in the '90s I spent filming a documentary about spinners, darkness marked the frustrating end of our workday, the time we were forced to leave the school behind, to listen wistfully to the sounds of their leaps and spins as they splashed on an ocean surface we could no longer see. They were racing offshore to begin diving into the deep scattering layer. This much we knew. But in filmmaking parlance, it was called "dip to black." Because what the dolphins did down there in the dark was unknown, and seemingly unknowable.

JUST ABOUT THE TIME WE drop anchor off Oahu, and unbeknownst to us, a catastrophe is being unleashed 4,400 miles and five time zones away, in the Gulf of Mexico. A mile below sea level, methane is shooting up the experimental well drilled by the *Deepwater Horizon* rig, exploding at the well's head, killing 11 workers, and igniting a firestorm. After 36 hours of a raging inferno—and still unknown to any of us—the rig will sink and open a valve to the gargantuan reservoir of the Macondo oil field, estimated to contain perhaps as much as 1 billion barrels, or 42 billion gallons, of crude.

Though it won't be understood for weeks, the *Deepwater Horizon* is different from any other spill in human history. The extreme technology used to drill at unprecedented depths lacks the extreme safety equipment and protocols needed to stave off disaster. BP, gambling at the border of controllable engineering, has lost spectacularly in its bid to be the deepest and cheapest driller of them all.

And no one is ready for it. Not the Minerals Management Service, catering submissively to BP's laughable Gulf oil-spill "plan," a document featuring wildly inaccurate wildlife assessments (including walruses and other species nonexistent in the Gulf) and an on-call expert who's been dead for years. Not the scientists whose research is paid for by the oil cowboys. Not the environmental groups, who did not foresee the stupendous potential for cataclysm on oil's farthest frontier. Not the media, who almost entirely ignored the sneak preview offered last year by the blowout of the West Atlas rig drilling in the Timor Sea off Australia—a disaster that required five attempts at a relief well and 74 days to stanch. Far offshore, far from sight, far beyond the typical royalty-paying boundaries, BP and its partners have transformed themselves into modern-day pirates, operating beyond law or conscience. Their reckless quest has endangered and perhaps condemned not just the Gulf Coast, but the largest, richest, most pristine, most biologically important, and last completely unprotected ecosystem left on Earth: the deep ocean.

Despite an ever-expanding estimate of the volume of the spill, relatively little oil washes ashore at first, and only a small portion ever will. Instead, trapped in the deep, the oil fouls the ocean's twilight and dark zones: the mesopelagic and the bathypelagic (bathos: deep). After April 20, the dumbwaiter rising through the waters of the Gulf of Mexico will be ascending an ocean fouled with a toxic broth of oil, methane, chemical dispersants, and drilling mud. The relatively small amounts of oil washing ashore, and the relief felt when the surface oil began to dissipate, hardly account for the devastation being wrought in the dark world beyond our sight.

SIX WEEKS AFTER THE *Deepwater Horizon* explosion, I'm aboard a small inflatable Greenpeace boat, bucking the marshy waters of Barataria Bay, Louisiana. A tide change is under way. Incoming and outgoing waters are flowing in opposing directions, battling each other in current lines inked with oil. A continuous flow of vessels chug through the pass—tugboats, barges, mud boats, seiners, trawlers, pirogues, airboats, sportfishers, pleasure cruisers. Some carry crews to and from the thousands of other drilling platforms puncturing the seafloor of the Gulf of Mexico, but the majority are now laden with containment boom and BP cleanup crews.

Dolphins are swimming in the pass too, a few dozen of an estimated 138 to 238 bottlenose dolphins that call Barataria Bay home. They're hugging the greasy waves of the tidal rip. Like bottlenose dolphins the world over, and like much marine life in general, they're exploiting the edge where waters of different provenance (temperature, salinity, velocity) hide predators from prey and vice versa. Along these edges, the sensory systems of the sea—sight, sound, pressure wave, magnetic field—are dimmed or distorted, making it difficult to see from one side through to the other. Bottlenose dolphins use the distortions as natural hunting blinds.

These waters have been off-limits to human fishers for weeks. But nobody told the dolphins. They're actively fishing the tidal rip and following trawlers dragging boom, because these are the same boats that sometimes give them food in the form of bycatch thrown overboard.

As best we know, the dolphins of Barataria Bay comprise a closed population whose members rarely if ever leave the bay. In theory, they could now exit, but in all likelihood they're trapped here by multiple barriers: by oily waters, by seasonal tradition, by cultural habit, by territorial boundaries, and by the availability of food—including fish and other marine life that may be trying to escape the oil by swimming inshore. At the moment, the dolphins are feeding as best they can in home waters that will likely kill them.

Rick Steiner, a conservation specialist from the University of Alaska who's studied the effects of the Exxon Valdez spill for the past 21 years, discusses these possibilities as we look on helplessly. "The dolphins aspirate oily fumes through their blowholes," he says. "They're eating fish exposed to oil. They're getting oil in all their orifices. They're bathed in a continual soup of oil. There's nowhere to go to get away from it. We know from the Exxon Valdez that even those animals not killed outright suffer lesions in their organs, including the brain. They go blind. They experience reproductive failures, changes in their blood chemistry, and possibly multigenerational changes passed down to offspring never even exposed to the oil."

A few hundred yards away, tucked into the marsh grass on Grand Isle State Park, we see a dead dolphin, half-skeletonized, half-mummified. In the heat and humidity of coastal Louisiana, it is hard to tell if it'd been dead a week or a month. We do know that dead dolphins are washing up along the Gulf Coast in higher-than-normal numbers. We don't know how many more have died at sea and sunk, never to be counted. On the beach surrounding the dead dolphin are hundreds of hermit crabs coated with a chocolatey syrup of oil, their tracks up the beach splattered as they fled the foul waters. The oil washing ashore is still actively bubbling. "Even though this concoction may have exploded from the well a month ago and has been wending its way ashore ever since, it's still full of volatile compounds like benzene," says Steiner. "Benzene's a known carcinogen, dangerous to human life, too."

Barataria Bay has become a hospice wilderness, full of dying plants and animals. Nearly all the marshy islands are oiled. The oyster beds covering 10 percent of the bay are dead or dying and now closed to human harvesting. The post-larval brown shrimp migrating into the bay (the estuaries of Louisiana and Texas are home to the highest densities of brown shrimp in US waters) are running an oily gauntlet. So are the speckled trout that normally feast on brown shrimp during their own breeding season. For the first time in my bird-watching life, I've seen multitudes of clapper rails—notoriously secretive marsh-dwelling birds—running down levees and roads in broad daylight trying to escape the oiled wetlands.

The fate of the marshes is inextricably linked to the fate of the deep ocean—and vice versa. The deep ocean seeds the marshes with the larvae of fish and invertebrates, which then repopulate the deep in their juvenile or adult stages. These inshore-offshore migrators include ecologically and commercially important species. Fifty percent of the wetlands in the lower 48 states line the Gulf of Mexico and produce more seafood than the Chesapeake Bay, South and Mid-Atlantic, and New England fisheries combined. Endangered Atlantic bluefin tuna, scheduled to spawn right now in the waters around the *Deepwater Horizon* blowout, migrate here because the Gulf's marshes—the ocean's womb—likely shelter and feed their larvae. Adult bluefin, deep divers, are hunting the depths to 3,300 feet in search of squid and crustaceans in the deep scattering layer. BP's oil will wallop them at all stages of their lives.

At Queen Bess Island, an important seabird rookery near the mouth of Barataria Bay, Steiner and I watch oily brown pelicans trying to preen themselves clean. I visited this same island a week ago; the downy pelican chicks who were still in the nest then are today slipping on oily rocks at the waterline. Where last week there were still a few dozen white pelicans, now there are only two, standing uncharacteristically alone, wings drooping in stress. Steiner points out the pelicans flying overhead, their bellies coated with oil. "Even those birds who are managing to avoid diving into contaminated water to feed are inadvertently floating on it," he says.

Death by oil is a horrible way to go. Necropsies on birds reveal hypothermia resulting from oiled feathers, malnutrition resulting from the hypothermia, anemia from the shock and stress of hunger, and poisoning from the oil ingested and inhaled during preening. Although a few birds will escape the immediate lethal effects, their eggs and chicks will not. An experiment from the 1980s with nesting Leach's storm-petrels—tiny seafaring birds breeding on islands off Newfoundland—found that birds exposed to crude oil or Corexit (the dispersant BP is using in the Gulf) lost more eggs and chicks than did control birds. This, even though the oil exposure was sublethal, and even if only one adult of the pair was oiled. Breeding success for adults generally returned to normal the following year—except in the case of birds exposed to the highest sublethal doses of oil or Corexit. Fewer of those birds returned to breed—indicating that their part in the experiment had proved lethal after all.

As bad as it is in Barataria Bay, it's only the beginning.

FROM THE OUTSET, BP HAS fought to control every aspect of its uncontrollable catastrophe other than the spill itself. It has wildly spun the numbers on the quantity of hemorrhaging oil. It has continued to dispense Corexit—above and below water—when ordered to stop. It has restricted press access with Kafkaesque flair. Unable or unwilling to skim much oil, BP has poured its energies into skimming up all available resources: renting virtually every hotel room on the Louisiana shores, helping to keep the press at bay; buying the silence of scientists with lucrative pay and confidentiality clauses; chartering nearly every boat on the coast and employing virtually every fisherman and captain made jobless by the spill. I find clusters of these men in the marshes and out in the Gulf, their boats tethered together so they can watch movies on the biggest boat's DVD player.

"They have to pay these guys to work or else they'll riot," says Carl Safina, marine conservationist and cofounder of the Blue Ocean Institute. "As it is, they're angry, drinking, griping in the bars. By paying them, BP is deflecting their anger. Plus some of them feel like they're really helping, even though BP's two prime cleanup methods—setting out boom and using dispersant—completely undermine each other."

The containment and absorbent boom that BP is deploying around beaches and marshes—largely ineffectively—is designed to do just that: contain and absorb oil. But the Corexit dispersant BP has flooded onto the leaking wellhead 5,000 feet down, and sprayed from the air onto the surface—some 2 million gallons in total—is designed to break up the oil. "Which one is it?" asks Safina. "Do you want to contain it or disperse it? It makes absolutely no sense to be doing both. Let's face it, with pollution, you count your lucky stars if you have what's called point-source pollution, that is, a single identifiable localized source of pollution, like the *Deepwater Horizon*. So what's BP doing with that? They're turning it into the worst pollution nightmare of them all: non-point-source pollution."

That's because untreated oil quickly rises to the surface, where it can be skimmed with relative ease. But treated with dispersant, it becomes a submerged plume, unlikely to ever float to the surface, and destined to migrate through underwater currents to the entire Gulf basin and eventually the North Atlantic. "Oil is toxic to most life," says Steiner. "And Corexit is toxic to most life. But the most toxic of all is oil that's been treated with Corexit. Plus, dispersants may well kill the ocean's first line of defense against oil: the natural microbes that break oil down for other microbes to eat." The EPA has never seriously examined Corexit's effects on marine life. Now it'll get the biggest and baddest field experiment of all time, as the flora and fauna of the shallows and the deep scattering layer collide with the dispersed plumes.

BP's schizophrenic approach to the cleanup becomes more insidious in light of the company's legal liabilities: The Clean Water Act stipulates that BP must pay $1,100 for every barrel of oil proven to have been spilled—$4,300 per barrel if gross negligence is determined. But the use of dispersants clouds estimates of the spill's size, guaranteeing that the true number will never be known—since relatively little oil will ever wash ashore—and guaranteeing that BP's liability will be vastly underestimated.

Consider that while we've all been fixated on the true spill rate—is it 35,000 barrels a day? 60,000 barrels? More?—those figures are only estimates, and only of the oil. Few people realize that some 40 percent of what spews from the *Deepwater Horizon* well is methane, the primary component of natural gas—a dangerous greenhouse gas and a toxin to most life. Indeed, methane may hold the answer to the quantity of vented oil. David Valentine, a biogeochemist at the University of California-Santa Barbara, suggested in May in an op-ed in the journal *Nature* that plumes of dissolved methane could be used to calculate how much oil has leaked into the Gulf of Mexico. But BP has blurred the evidence trail—intentionally or otherwise—by treating at least some of the escaping methane with methanol, another toxin, in an effort to prevent a dangerous buildup and possibly even another explosion. Nevertheless, around the spill site, Valentine and his colleagues found clouds of dissolved natural gas at 100,000 times the normal density and at depths of more than 2,500 feet. They also found that little of the gas seemed to be reaching the air. Which is good news for the atmosphere, but probably bad news for the ocean. That's because the methane may also be powering up blooms of microbes that eat methane but use up the oxygen in the water as they do so—causing dead zones where most life cannot survive. The Gulf of Mexico is already home to the second-largest dead zone on Earth; the last thing it needs is another. On the surface above the methane clouds, Valentine and colleagues discovered a mass kill of pyrosomes—free-floating colonies of jellyfish look-alikes that straddle the vertebrate-invertebrate divide, and an important food for sea turtles. It's not yet clear which of many smoking guns killed the pyrosomes. "We'll be working up the story of the relationship between dispersant, oil, gas, and the microbial community for some time to come," says Valentine.

Then there are the drilling fluids contaminating the seafloor near the wellhead. Euphemistically called muds, these heavy fluids are pumped into wells to keep the highly pressurized oil and gas from exploding upward. BP's drilling muds have been pouring out of the wellhead, along with 30,000 barrels added in its failed "top kill" and other efforts to plug the leak. Along with oil, methane, methanol, and Corexit, drilling fluids add their own frightening recipe to the disaster: arsenic, lead, mercury, cadmium, barite, fluoride, chrome lignosulfonate, vanadium, copper, aluminum, chromium, zinc, radionuclides, and other heavy metals. Relief wells require pumping thousands more barrels of drilling fluid into the reservoir, with all the same risks of explosion attending the original well. The EVA [Environmental Vulnerability Assessment] estimates these drilling fluids will pose a threat to the seafloor and surrounding waters for up to 40 years. Plus a recent study finds that oil spills create a whole new pathway for arsenic pollution in the sea. The oil prevents seafloor sediments from bonding with and burying arsenic that naturally occurs in the ocean. This shutdown of the natural filtration system allows arsenic levels to rise from the deep water to the surface, disrupting photosynthesis in phytoplankton, increasing birth defects and triggering behavioral changes in marine life, and killing animals that feed on poisoned prey.

Never before in human history has the vast food web of the ocean—rooted in the dark, and flowering at the surface—come under so many assaults from below, above, and within the water column: marine warfare masquerading as a cleanup.

"WE DROVE DOWN IN CLEAR water but came up 30 minutes later through oil," says Nancy Rabalais, director of the Louisiana Universities Marine Consortium (LUMCON), a research station tucked deep in the marshes of southern Louisiana in the village of Cocodrie. A few weeks after the spill, during her summer research surveys 10 miles offshore, Rabalais personally encountered BP's plumes, which will probably affect her research far into the future. "It was horrible," she says, grimacing. "We were covered. Our gear was covered. We were breathing fumes and tasting oil."

The last time I saw Rabalais, after Hurricane Katrina in 2005, LUMCON was trashed: the station evacuated, the marshes littered with drowned trees, broken boats, unroofed houses. The area is mined in a whole new way today. Along with the oil, dispersant, benzene, and everything else creeping into the bayous, Cocodrie has become a staging point for BP—complete with Louisiana National Guard troops, workers recruited from all over the South, and fishermen hired away from their extinct jobs. These men are cashing in their lunch chits at the Coco Marina restaurant, where Rabalais, Ed Chesney—LUMCON'S fisheries biologist—and I are grabbing a meal. We watch every manner of boat known to Louisiana speed up the narrow channels to the marina, their white hulls stained BP brown, their wakes slapping the cordgrass flat. The boats offload hundreds of hungry men.

Rabalais is worried about the species already under enormous stress from a host of other environmental problems in the Gulf: dead zones, overfishing, chronic oil pollution, seismic testing for oil and natural gas, coastal erosion. "Brown pelicans just came off the endangered species list," she says, "and now some of their most important breeding rookeries are getting hit with oil." She's concerned about critically endangered Kemp's Ridley sea turtles, the rarest on Earth, a species that faced mortal threat from the 140 million-gallon spill at the Ixtoc I drilling platform in the Gulf in 1979. Kemp's Ridleys breed almost exclusively in the Gulf, with virtually every female returning to lay her eggs on a stretch of beach south of the Texas border.

In the wake of the BP spill, there's been a spike in sea turtle deaths, the majority of them Kemp's Ridleys. The number is certain to rise, since some sea turtles feed in the DSL, and most enjoy a meal of jellyfish. Sadly, they also eat blobs of oil they mistake for jellyfish. According to some reports, sea turtles have been roasted alive in the surface-oil patches burning offshore. Hundreds more have drowned since the disaster began. One shrimp fisherman privately admits that panicky colleagues fished hard in the weeks after the spill, knowing that the fishery would soon be closed, and some tied shut the mandatory turtle-excluder devices, which save turtles from drowning but reduce the efficiency of their nets.

Rabalais and others also worry about the Gulf's sperm whales, which feed on squid living in the deep scattering layer. An estimated 1,665 sperm whales inhabit (and perhaps never leave) the northern waters of the Gulf. A recent National Oceanic and Atmospheric Administration (NOAA) assessment calculated that even three additional deaths (by other than natural causes) could endanger the entire sperm whale population, since the whales breed infrequently and only in midlife. The whales favor the deep waters of Mississippi Canyon—the location of the *Deepwater Horizon* wellhead. On numerous occasions, they've been seen swimming through thick oil in that region. And it's not only sperm whales. The Gulf is home to 29 species of cetaceans, many of which feed on the DSL, including spinner dolphins, spotted dolphins, pilot whales, killer whales, and many secretive deep divers such as beaked and bottlenose whales. The filter-feeding whales—including the Gulf's tiny isolated population of Bryde's whales, plus humpbacks, fins, minkes, and sei, many of which are DSL feeders—are vulnerable a whole different way, since oil fouls their baleen (sievelike teeth), dooming them to starvation.

And then there are the 400 Florida manatees, a species classified as vulnerable to extinction, that migrate to Louisiana waters each summer. This year they'll be feeding in oily water on oiled algae and cordgrass. "And it's not just the large fauna we worry about," says Rabalais. "The entire wetland is at risk. A marsh that's been heavily oiled becomes anaerobic at the roots. The next time a big storm comes through, those marshy islands will in all likelihood just break up and disappear." If so, they'll take the nursery grounds for marine life with them. Coastal Louisiana is already losing 24 square miles of wetlands a year, a football field every 30 minutes. These dwindling wetlands are crucial to the Louisiana economy, keeping people here afloat in businesses from fishing to tourism. "Now they're all out of work," says Rabalais. "And the revenues we were counting on to re-build the coastal habitats to foster the birds, shrimp, fish, dolphins, turtles, whales, and people will be lost."

Least certain of all is what's happening to the life at the bottom of the Gulf of Mexico. Take the life that congregates around cold methane seeps, the first of which ever discovered was found in the Gulf in 1984. Since then, 50 more sites have been located in these waters, some close to the *Deepwater Horizon,* with hundreds more likely out there—all home to otherworldly collections of crustaceans, snails, bacterial filaments, and tubeworms. The rules of life are different in the gassy depths, where life capitalizes on the same fossil fuels we're drilling for. Some cold-seep tubeworms have lifespans of 250 years. Others recently found in the deepest seeps may live to 500 or 600 years.

Though some of these creatures feed on methane, that doesn't mean they can survive the spill. "The quantity of oil and the added effects of dispersants are likely to harm these communities," says

Lisa Levin, a biological oceanographer and cold-seeps specialist from the Scripps Institution of Oceanography. Oil could smother the animals' feeding apparatus or suffocate the bacteria at the base of the food chain, she adds. "The tubeworms and other seep organisms, including perhaps deepwater corals, are so slow-growing that damage will likely be long lasting." Levin envisions a host of long-term chronic problems throughout the deep Gulf that might not even show up for decades.

Only 25 miles from the *Deepwater Horizon* blowout, a tremendously rich area known as the Pinnacles hosts deepwater corals 300 to 500 feet below the surface. One of the Gulf's invisible splendors, these ancient reefs line the outer continental shelf south of Mississippi and Alabama. During the last ice age, when today's continental shelves were dry land, the Pinnacles were living coral reefs near the shoreline. Nowadays the fossil reefs lie too deep and dark for most reef-building corals or phytoplankton to survive. Instead, they're largely fueled by zooplankton, which power rich deepwater communities of soft corals, sponges, feather stars, black corals, solitary hard corals, and predatory fish, including reef fish not found in shallow waters. The site is also a critical spawning habitat for commercially important species like grouper and snapper.

"We lack even a good picture of life in the deep Gulf," says Ed Chesney. "Now we may never know what's been done to it." It's the classic iceberg equation: a nine-tenths submerged hazard, lurking unseen in the darkness. The big question: Will it wreck the Gulf of Mexico? "The best thing that might happen now," says Chesney, a battle-scarred veteran of Hurricanes Katrina, Rita, Gustav, and Ike, "is for one, two, three, or four hurricanes to blow through and bury all this pollution under layers of sediment."

His thinking is that the tons of silt accompanying storm surges would inter the contamination and prevent it from migrating further, while more silt stirred up offshore would provide particles for the emulsified oil droplets to adhere to and sink to the bottom.

Huge offshore waves could also trigger subsurface landslides to bury some of the polluted seafloor under clean sediment: nature's dip to black.

Yet the potential benefits of hurricanes are accompanied by obvious risks. Hurricanes will drive pollution farther inland. The 33,000 miles of pipeline in the Gulf's waters and marshes are critically vulnerable to hurricane-induced waves. Seven weeks after the *Deepwater Horizon* spill, naval scientists released the results of research conducted when Hurricane Ivan swept through the Gulf in 2004. It found these pipelines to be far more vulnerable than previously thought to deep storm currents, which slosh for up to a week with enough force to break pipelines 300 feet deep. Plus every hurricane in this storm-prone region threatens the cement seals on 50,000 holes punched into the floor of the Gulf: some wells in deep water, many in the shallows, 27,000 of them abandoned and unmonitored, 600 once run by BP. The passage of Katrina spilled 8 million gallons of oil from platforms, pipelines, ships, and storage tanks—three-quarters as much oil as was dumped by the Exxon Valdez.

All of which adds up to the realization that our collective "don't ask don't tell" attitude toward the deep ocean—mining it, drilling it, overfishing it, dumping in it (including nuclear waste), polluting it, and deafening and killing its life with lethal sounds produced by the drilling industry and the military—is a prescription for ruin. "It's not that we were totally unprepared for the possibility of the *Deepwater Horizon*," says Carl Safina, "but that we were so spectacularly unprepared for its inevitability."

IRONICALLY, THE TOOLS KELLY BENOIT-BIRD and Margaret McManus are employing in Hawaii to decipher the deep scattering layer were developed by the offshore oil and gas industry and the military. "The DSL was a hot topic during the Cold War," says Benoit-Bird, "but only its acoustic properties, not its biological properties. American and Soviet navies wanted to know how to use its sound-reflecting properties to hide their submarines." In the 21st century, the application has shifted to the oil industry's fight to drill deeper, a battle spurring technological innovation in echo sounding and imaging equipment—including the "spill-cam," whose footage BP was finally pressured into releasing. "As offshore rigs proliferate and get deeper," says Benoit-Bird, "the once-prohibitively expensive gear attending them became cheaper and more accessible, to the point where the smallest players, the research scientists like Margaret and me, can now afford some of it."

The data streaming in from the waters off Oahu—the yellow, green, red, and blue bands scrolling across the computer monitors— are unprocessed data, designed to signal that the submerged gear is working correctly. Back in Benoit-Bird's office at Oregon State University, I watch the information transform into geek IMAX. The animations show spinner dolphins gathering in a circle of 16 to 28 animals, always an even number, each dolphin paired with another, the pairs arranged in an echelon formation: one animal slightly above and ahead of the next, while maintaining about three feet of separation. A perimeter of roughly 300 feet is precisely maintained as the dolphins swim in an undulating circle, trapping the fish inside the net of their swimming bodies.

One after another, in fixed sequence, two dolphin pairs directly opposite each other dart into the ball of fish to feed. As they return to the circle, four more follow. And so on. The action is extremely fast, the dolphins darting in to feed at a rate of roughly 1.25 prey per minute, all while swimming and circling in their roller-coaster pattern. After five minutes below, each pair has engaged in four feeding dashes, and the dolphins simultaneously surface to breathe. They typically grab only one or two quick breaths before diving, repeating the underwater rodeo over and over throughout the night without rest. "Our research indicates that spinner dolphins are forced to fish hard and continually all night," says Benoit-Bird, "and to catch the biggest of these tiny four-inch-long fish they possibly can in order to meet their metabolic requirements."

In other words, they exploit a different kind of edge—the fine line between survival and starvation. This precarious balance tips back and forth across the food web of the deep scattering layer. "In order to really understand what the dolphins are doing," says Benoit-Bird, "we had to understand what their prey are doing. And in order to do that, we had to decipher what's behind the movements of the deep scattering layer. This investigation led us incrementally backward over time towards the smaller and smaller organisms—which, as it turns out, drive the entire system."

Margaret McManus was part of the team that first discovered a remarkable phenomenon rewriting our understanding of ocean dynamics—the formation of thin plankton layers in the ocean. These congregations of plankton, both the plant and animal varieties, may extend for many miles horizontally but inhabit a few feet on the vertical scale—sheets of life packed far, far more densely with life than the water just above or below them. The formation of thin layers is driven by the chemistry and physics of the ocean, as well as by the organisms themselves. Off Hawaii, they tend to form where cooler waters well up from the deep during tide changes.

McManus and Benoit-Bird have found that DSL fish will swim hard against prevailing currents in order to get to these dense aggregations of life. It's an energy-consuming choice offset by the rich feeding rewards. In Benoit-Bird's data animations, single fish dive into a thin zooplankton layer and swim up and down, back and forth, eating a doughnut hole in the layer. The spinner dolphins do something similar: diving to find patches of lantern fish that they then corral increasing the prey density by up to 200 times. "It's so congested in there for these nonschooling fish of the DSL," says Benoit-Bird, "that they're probably bumping into each other in confusion."

The emerging picture is one of an incalculably complex, finely tuned, and delicate interaction between predators and prey, chemistry and light, currents and water column, night and day. Some semblance of this spatial ballet, played in weightless three-dimensional darkness, has likely been part of the oceans since the oceans were brought to life: layers of life gathering in extremely high densities to feed or to avoid being eaten.

So what happens if you add millions of gallons of oil, dispersant, methane, and drilling fluid into the dense mix?

"We know that the deep scattering layer in the Gulf of Mexico— like the DSL everywhere—supports huge numbers and biomass of life," says Benoit-Bird, who has spent time studying the Gulf's sperm whales. "We know the DSL is super important to the life of those waters. We know it's constantly on the move, not only up and down, but inshore and offshore, back and forth, every day and every night. This greatly increases the likelihood that any given animal or layers of life will be exposed to the pollutants at some point in the course of their travels. And each of these exposures will cascade up and down through the food web."

Some early observations of the effects of the Gulf catastrophe suggest the daily vertical migrations of the animals of the deep scattering layer may be blocked when they encounter plumes of oil and contaminants. If so, then trapped below a plume, the DSL fish and invertebrates would be unable to access their prey. Trapped above, they would be unable to escape their predators. Trapped within, they would probably die—and in their deaths, poison those who eat them. For the ocean, any loss of productivity in the deep scattering layer would be the biggest cataclysm of all—impoverishing the surface waters, depleting the coasts, cascading across the boundaries between ocean and land to denude both natural and human economies.

BEFORE BEING WAYLAID BY THE oil tragedy, I was investigating the emergence of a better future for the ocean—one in which we could use our scientific and technological genius to create a new, exciting, and profitable relationship with our water world, a relationship based on respect and sustainability. I spent a few weeks in Hawaii, where the larvae of many promising ideas are circulating on scholarly and entrepreneurial currents.

At the University of Hawaii-Manoa, I met Luis Vega, who drifted years ago from his natal shores of Peru and landed in American academia. His shock of white hair and his melancholic, ironic air give him the guise of a poet. He told me that when he was working on clean energy in the Jimmy Carter years, he was a popular man. Then he weathered decades of solitude. "Now I'm popular again," he smiles self-deprecatingly.

Vega is one of the foremost modern developers of OTEC (ocean thermal energy conversion) technology. He managed the design, construction, and operation of an experimental OTEC plant for the production of electricity at the National Energy Laboratory of Hawaii Authority (NELHA) on the Big Island from 1993 to 1998. Today Vega has a new grant, via the Department of Energy and Lockheed Martin, to essentially see if the technology is suitable for commercial investment. "Today, while we talk about wind, solar, and wave power," says Vega, "we're ignoring this energy inherent in the ocean, a source far more powerful, far more consistent, than any of those. The beauty of OTEC lies in its unshakable ability to provide energy 24/7, without any of the vagaries of wind, solar, or wave."

OTEC runs on the temperature differential between the ocean's deep dark waters and its warmer sunlit zone—the same differentials the creatures of the DSL exploit. In a closed OTEC system, warm surface waters are pumped through a heat exchanger to vaporize a low-boiling-point fluid, like ammonia. Cold deep seawater is simultaneously pumped through a second heat exchanger, creating a gradient that drives the vapor through a turbine to generate electricity.

Finally, the cold seawater condenses the ammonia back into a liquid, to be recycled through the system. Both Japan and India are also experimenting with OTEC power plants.

OTEC isn't the deep water's only use. At NELHA, the two cold-seawater pipes built for the last OTEC experimental plant today deliver water from between 2,000 and 3,000 feet deep to dozens of surrounding businesses. The 43-degree, extremely clean water enables aquaculture farms to grow cold-water seafood like Japanese abalone, flounder, oysters, and Atlantic lobster in the tropics. The deep water is also being used to raise aquarium fish and helps grow spirulina at one of the largest algae farms on Earth. The hope is that these methods could one day offer a sustainable alternative to wild-caught fish, especially disappearing species, like tuna.

But generating scalable commercial power would clearly be the killer app. The way Vega envisions our energy future, the first generation of OTEC "plant-ships" would be stationed offshore and send electricity via subsurface power cables to shore stations. Then, in 20 or 30 years, the technology would develop to the point where "grazing" OTEC plants could decouple from the land and roam tropical waters in search of the best temperature differentials. These second-generation plant-ships would exploit those differentials, using the energy to break down seawater and create energy-rich compounds like hydrogen or ammonia. Either could essentially serve as a battery—holding energy as it's transferred to land. And in the case of hydrogen, there might be a robust infrastructure in place to distribute liquid hydrogen (such an infrastructure is being built in California) to be used in fuel-cell vehicles.

Figuring out how or whether OTEC or any of these other alternate energy technologies can provide us with a livable future will take serious investment. Yet until now we've barely acknowledged the true costs of subsidizing "cheap" oil: not only the $4 billion a year in actual subsidies, but climate change, the risks to human health, environmental degradation, and disaster. In the wake of the

Gulf of Mexico tragedy, alternative energy sources—OTEC, wave, tide, wind, or solar—no longer seem utopian, merely sane.

On the Big Island of Hawaii, the NELHA deepwater pipes run up near a beach park on the shoreline. In the early morning, I see a school of spinner dolphins in the blue water just beyond the breakers. Their night's intensive work finished, they're leaping and spinning their way back to shore. During my years filming spinner dolphins, I sometimes joined them underwater during their morning return to land. The sight was a marvel of speed and grace, dozens of slender bodies streaming below the surface at velocities that transformed the school into a waving, blurry contrail of gray and white and black. For me, stationary in the water while the spinners streamed past, as the sun ignited the twilight water, it felt like being inside the eye of a hurricane of intensive, productive, pure energy.

On the shores of the Gulf of Mexico, as black doom wells up from the seafloor a mile down, I find oil on beaches repeatedly cleaned by hazmat crews. All I have to do is lean down and scratch an inch into the sand to find goop. It occurs to me that a new stratum is being written in the geological logbook of the Gulf of Mexico, perhaps someday to be known as the BP dark layer. Will history record it as the oily seam marking the end of an untenable energy era and the beginning of a better one?

A dip to black isn't always the end of the story. Sometimes it's followed by a fade up from black and a whole new scene.

CYNTHIA GORNEY

The Estrogen Dilemma

FROM *THE NEW YORK TIMES MAGAZINE*

> *When the results of the Women's Health Initiative indicated that hormone-replacement therapy created a higher risk for heart disease, stroke, and breast cancer, many women—and their doctors—found themselves at a loss to know how to treat the sometimes brutal symptoms of menopause. As Cynthia Gorney has learned, researchers are finding that there is more to the estrogen story than originally thought.*

HERE WE ARE, TWO FAST-TALKING WOMEN ON estrogen, staring at a wall of live mitochondria from the brain of a rat. Mitochondria are cellular energy generators of unfathomably tiny size, but these are vivid and big because they were hit with dye in a petri dish and enlarged for projection purposes.

They're winking and zooming, like shooting stars. "Oh, my God," Roberta Diaz Brinton said. "Look at that one. I love these. I love shooting mitochondria."

Brinton is a brain scientist. Estrogen, particularly in its relationship to the health of the brain, is her obsession. At present it is mine too, but for more selfish reasons. We're inside a darkened lab room in a research facility at the University of Southern California, where Brinton works. We are both in our 50s. I use estrogen, by means of a small oval patch that adheres to my skin, because of something that began happening to me nine years ago—to my brain, as a matter of fact. Brinton uses estrogen and spends her work hours experimenting with it because of her own brain and also that of a woman whose name, Brinton will say, was Dr. A. She's dead now, this Dr. A. But during the closing years of her life she had Alzheimer's, and Brinton would visit her in the hospital. Dr. A. was a distinguished psychotherapist and had vivid stories she could still call to mind about her years in Vienna amid the great European psychologists. "We'd spend hours, me listening to her stories, and I'd walk out of the room," Brinton told me. "Thirty seconds later, I'd walk back in. I'd say, 'Dr. A., do you remember me?' And she was so lovely. She'd say: 'I'm so sorry. Should I?'"

The problem with the estrogen question in the year 2010 is that you set out one day to ask it in what sounds like a straightforward way—Yes or no? Do I or do I not go on sticking these patches on my back? Is hormone replacement as dangerous in the long term as people say it is?—and before long, warring medical articles are piling up, researchers are raising their voices and gesticulating excitedly and eventually you're in Los Angeles staring at a fluorescent rodent brain in the dark. "You want a statistic?" Brinton asked softly. Something about the shooting mitochondria has made us reverent. "Sixty-eight percent of all victims of Alzheimer's are women. Is it just because they live longer? Let's say it is, for purposes of discussion. Let's say it's just because these ladies get old. Do we just say, 'Who cares?' and move them into a nursing home? Or alternatively, maybe they are telling us something."

With their brains, she means. Their sputtering, fading Alzheimer's brains, which a few decades earlier were maybe healthy brains that might have been protected from eventual damage if those women had taken estrogen, and taken it before they were long past their menopause, while their own neural matter still looked as vigorous as those rat cells on the wall. This proposition, that estrogen's effects on our minds and our bodies may depend heavily upon when we first start taking it, is a controversial and very big idea. It has a working nickname: "the timing hypothesis." Alzheimer's is only one part of it. Because the timing hypothesis adds another layer of complication to the current conventional wisdom on hormone replacement, it has implications for heart disease, bone disease and the way all of us women now under 60 or so—the whole junior half of the baby boomers, that is, and all our younger sisters—could end up re-examining, again, everything the last decade was supposed to have taught us about the wisdom of taking hormones.

I first met Brinton at a scientific symposium at Stanford University in January that was entirely devoted to the timing hypothesis. The meeting was called Window of Opportunity of Estrogen Therapy for Neuroprotection, and it drew research scientists and physicians from all over the country. When I asked to listen in, the organizers hesitated; these are colleagues around a conference table, they pointed out. They're probing, interrogating, poking holes in one another's work in progress.

But I was finally permitted to take a chair in a corner, and as the day went on, I became aware of my patch, in a distracted, hallucinatory sort of way, as if I had started fixating on a smallish scar. One after another, their notes and empty coffee cups piling up around them, heart experts and brain experts and mood experts got up to talk about estrogen—experiments, clashing data, suppositions, mysteries. There are new hormone trials under way that are aimed at the 40-year-old to 60-year-old cohort, with first results due in 2012 and 2013. There are depression studies involving estrogen. There are dementia studies involving estrogen. There are menopausal lab monkeys taking estrogen, ovariectomized lab mice taking estrogen and young volunteers undergoing pharmaceutically induced menopause so researchers at the National Institutes of Health can study exactly what happens when the women's estrogen and progesterone are then cranked back up. I typed notes into my laptop for hours, imagining the patch easing its molecules into the skin of my back, and the whole time I was typing, working hard to follow the large estrogen-replacement thoughts of the scientists around the table, I had one small but persistent estrogen-replacement thought of my own: If I make the wrong decision about this, I am so screwed.

I STARTED TAKING ESTROGEN BECAUSE I was under the impression that I was going crazy, which turns out to be not as unusual a reaction to midlife hormonal upheaval as I thought. This was in 2001. The year is significant, because the prevailing belief about hormone replacement in 2001 was still, as it had been for a quarter century, the distillation of extensive medical and pharmaceutical-company instruction: that once women start losing estrogen, taking replacement hormones protects against heart disease, cures hot flashes, keeps the bones strong, has happy effects on the skin and sex life and carries a breast-cancer risk that's worth considering but not worrying about too much, absent some personal history of breast cancer or a history of breast cancer in the immediate family.

At first, as I was trying to locate a psychiatrist who would take me on, I wasn't aware I had reason to pay attention to advice about hormones at all. That year I turned 47, a normal age for beginning the drawn-out hormonal-confusion period called perimenopause, but I had none of the familiar signs. Menopausal holdouts run in the family; one of my grandmothers was nearly 60 by the time hers finally kicked in. My only problem was a new tendency to wake up some mornings with a great dark weight shoving my shoulders toward the floor and causing me to weep inside my car and basically haul myself around as if it were the world's biggest effort to stand up straight and carry on a conversation. Except for its having shown up so arbitrarily and then coming and going in waves, there was nothing interesting about my version of what my husband and I came to think of as the Pit; anybody who has been through a depression knows what a stretch of semidisabling despair feels like, and for my part I had a very nice life, a terrific family and a personal interior chorus of quarreling voices demanding to know why I didn't pull up my socks and carry on, which in fact was the first question I planned to ask a psychiatrist.

But I went to my gynecologist first, so she could check my blood pressure or whatever seemed the prepsychiatrist thing to do. How often would you say you feel this way, she asked; and I said I didn't know, maybe every few weeks; and she told me to start keeping records. Note each day, she said. Check for patterns.

She was right. There was a pattern. I was falling into the Pit on schedule, around 11 days before each menstrual period, or M.P., which is one of many abbreviations I was to learn in my efforts to keep track of the ferocious hormones debate that started up in North America in 2002, one year after I stuck on the first estrogen patch that my gynecologist prescribed. The study at the center of the ruckus was called the Women's Health Initiative, or W.H.I. It was a federally financed examination of adult women's health, extraordinary in scale and ambition, that started up in the early 1990s; one of its drug trials enrolled more than 16,000 women for a multiyear comparison of hormone pills versus placebos. On July 9, 2002, W.H.I. investigators announced that they had ended the trial three years early, because they were persuaded that it was dangerous to the hormone-taking participants to let them continue.

The women on hormones were having more heart trouble than their placebo-taking counterparts, the investigators said, not less. Their risk for stroke went up. Their risk for blood clots went up. Their risk for breast cancer increased by 24 percent. The W.H.I. bulletins dominated medical news all summer and long into the fall, and so alarming were their broad-scale warnings that millions of women, myself included, gave up hormone replacement and resolved to forge ahead without it.

The patches my gynecologist prescribed worked, by the way. I didn't understand how, beyond the evident quieting of some vicious recurring hormonal hiccup, and neither did the gynecologist. But she had other women who came in sounding like me and then felt better on estrogen, and I would guess many of them, too, decided after the W.H.I. news that they could surely find other ways to manage their "mood swings," to use the wondrously bland phrasing of the medical texts. (I'm sorry, but only someone who has never experienced one could describe a day of "I would stab everyone I know with a fork if only I could stop weeping long enough to get out of this car" as a "mood swing.") We muddled along patchless, my mood swings and my patient family and I, until there came a time in 2006 when in the midst of some work stress, intense but not unfamiliar, I found myself in a particularly bad Pit episode and this time unable to pull out.

It was profoundly scary. In retrospect, I managed a surprising level of public discretion about what was going on; competence at the cover act is a skill commonly acquired by midlife women, I think, especially those with children and work lives. If the years have taught us nothing else, they have taught us how to do a half dozen things at once, at least a couple of them decently well. Like other women I have met recently with stories like this one, I relied for a few months on locked office doors, emergency midday face-washings and frequent visits to an increasingly concerned talk therapist. But one afternoon I got off my bicycle in the middle of a ride with my husband, because I had been crying so hard that I couldn't see the lane lines, and I sat down on the sidewalk and told him how much I had come to hate knowing that family obligations meant I wasn't allowed to end my life. The urgent-care people at my health clinic arranged a psychiatric consult fast, and after listening and nodding and grabbing scratch paper to draw me an explanatory graph with overlapping lines that peaked and plunged, the psychiatrist wrote me two prescriptions. One was for an antidepressant.

The other—I recognized the name as soon as she wrote it down—was for Climara, my old estrogen patch.

By this time we were four years past the 2002 W.H.I. hormone news. So I knew a few more things. I knew there had been a surge of industrious scrambling among former hormone-taking women, some of whom had tried multiple alternatives or going cold turkey and then changed their minds and re-upped on estrogen, deciding that life without it was so unpleasant that they no longer cared what the statistical prognoses said. I knew the prevailing medical sentiment had shifted slightly since the bombshell of 2002; certain articles and books still urged women to shun hormone replacement at all costs, but the more typical revised counsel was, essentially, proceed with great caution. If some menopausal malady is genuinely making you miserable, the new conventional wisdom advised, and no alternative remedy is working for you, then go ahead and take hormones—but keep the dose low and stop them as soon as possible.

I would like to be able to tell you that I weighed these matters thoughtfully, comparing my risks and benefits and bearing in mind the daunting influence of a drug industry that stands to profit handsomely from the medicalizing of normal female aging. But that would be nonsense, of course. I was too crazy. I went straight to the pharmacy and took everything they gave me.

You don't read the fine print on package labels when you're being ushered through a psychiatric crisis, but after a while, I did. By last winter I was nearing the cumulative five-year mark as an estrogen user, and although "low dose, stop soon" is often an advisory without specifics attached, five years seemed to turn up here and there as an informal outer-limit guideline. And because it had worked again, because the estrogen so clearly helped repair something that was breaking (there's no way for me to separate the effects of estrogen from the effects of the antidepressant, except that on the few occasions when I've been haphazard about replacing the estrogen patches on time, I've experienced prompt and unmistakable intimations of oncoming Pit), I now had some rational faculties with which to go looking for explanations that might help me decide what to do. This was when I first began learning that in the controversy over hormone replacement, the fine print matters a very great deal.

First of all, the kind of estrogen in my patches—there are different forms of estrogenic molecules—is called estradiol. It's not the estrogen used in the W.H.I. study. Pharmaceutical estradiol like mine comes from plants whose molecules have been tweaked in labs until they are atom for atom identical to human estradiol, the most prominent of the estrogens premenopausal women produce naturally on their own. The W.H.I. estrogen, by contrast, was a concentrated soup of a pill that is manufactured from the urine of pregnant mares. The drug company Wyeth (now owned by Pfizer) sells it in two patented products, the pills Premarin and Prempro, and it's commonly referred to as "conjugated equine estrogens."

There was more in the fine print. Two years ago, after warning me that women who haven't had a hysterectomy run a higher risk of uterine cancer when they take only estrogen as hormone replacement, a new doctor added in progesterone, which has been shown to protect the uterus. The progesterone he prescribed for me, like the estradiol, is a molecular replica of the progesterone women make naturally. It's different from the progesteronelike synthetic hormone that was used for the W.H.I. study that ended in 2002. That medication was a formulation whose multisyllabic chemical name shortens to MPA and which has a problematic back story of its own: MPA takes care of the uterine-cancer risk, but there's reason to suspect it may be a factor in promoting breast cancer. And it's ingested as a pill, which means that like equine estrogens (and unlike, for example, my patch), MPA metabolizes through the liver, possibly creating additional complications en route, before going about its business.

The biggest difference between me and the W.H.I. women, though, has to do with age and timing. I started on the patches while my own estrogen, pernicious though its spikes and plummets may have been, was still floating around at more or less full strength. The average age of the W.H.I. women was just over 63, though the study accepted women as young as 50. More significant, though, most of them were many years past their final menstrual period, which is the technical definition of menopause, when they began their trial hormones. The bulk of the group was at least 10 years past; factoring in the oldest women, the average number of years between the volunteers' menopause and their start on the trial medications was 13.4.

Because women generally make decisions about hormones while they are in the throes of perimenopause—that term is now used to extend through the year following the final M.P.—you may find this as perplexing as I did. Why would the largest drug trial in the history of women's health select, for most of its participants, women already long past the critical phase? I heard one undiplomatic critic sum up the W.H.I. as "the wrong drugs, tested on the wrong population," and those two factors, the drugs and the population, are actually directly linked. Equine estrogens and MPA were the only forms of hormones used in the W.H.I. trials. Among other reasons, that's because drug trials are expensive; this one was huge, and Wyeth was going to provide without cost an average of eight years' worth of its equine estrogens and MPA to 40 clinical centers.

And millions of women were using those very hormones already, partly because aggressive Wyeth marketing had for three decades insisted that hormone replacement was the ticket to a vigorous and sexually satisfactory postmenopausal life. To a certain extent, evidence backed up that claim; wide-scale though less rigorous earlier studies appeared to demonstrate hormone replacement's benefits so clearly that many physicians were suggesting it almost automatically to midlife women, whether or not they had perimenopausal complaints. Hormones raised the breast-cancer risk in those earlier studies, but nearly every other health factor showed improvement when women who took hormones were compared with those who didn't. Hot flashes disappeared, osteoporosis was milder, women reported feeling better and women who took hormones showed a markedly lower rate of heart disease than women who did not.

Because heart disease ultimately kills many more women than all cancers combined, some doctors had also taken to urging older women, even those past menopause, to start hormones for cardiac-health purposes. The W.H.I. trials were supposed to provide conclusive evidence, finally, as to whether all this wide-scale prescribing was truly a sound idea. But cardiovascular disease tends to make its bids for attention—its "events," as clinicians say, like heart attack and death—when we're quite a bit past 51, the average age at which American women hit menopause. The only way the W.H.I. was going to tally up a scientifically useful number of cardiac events was to enroll plenty of women already old enough to reach that danger stage before the study's time ran out. So that's what they did, and once the final data was reparsed many times, it was clear that the trial had shown physicians something highly important about the perils of starting older postmenopausal women (that's qualifier No. 1) on pills (No. 2) containing equine estrogens (No. 3) plus MPA (No. 4).

Those four qualifiers make the chief message of the W.H.I.—that taking hormones, in the long run, is more likely to hurt you than help—far more specific than the one most women heard. For those of us not yet on the far side of menopause, or who don't match the other qualifiers (as I write this, for example, I'm zero for four), a daunting proportion of what we thought we learned about hormone

replacement over the last eight years remains unsettled, more con-
fusing than ever and conceivably—we don't know yet—wrong. "I
mean, if you're a 70-year-old," says S. Mitchell Harman, a Phoenix-
based endocrinologist and coordinator of one of the national trials
currently examining hormones' effects on younger women, "and
your question is, Should I start taking estrogen? the W.H.I. answered
that for you beautifully. No. Unfortunately, it wasn't designed to
answer that question for a 50-year-old. So now we're trying to fill in
the blanks."

ONE AFTERNOON LAST MONTH, I reported to the Northern
California site for an N.I.H.-financed cognitive trial that is part of
the Kronos Early Estrogen Prevention Study that Harman is leading.
Keeps, as it's called, has enrolled women at nine such sites around
the country; this one was inside a medical building at the University
of California, San Francisco, and the cognition test I asked to try
proved to be a low-tech experience: a table with chairs, pens and
pencils and a gentle-voiced psychologist asking me to do things with
my brain. Number sequences repeated backward, lists of random
objects to recall, designs to remember and copy—I promised not to
describe specifics, because making details public could compromise
the trial results. But imagine a stranger holding up a stopwatch and
giving you 30 seconds to name every dessert item you can think of.
The brain charges off into a comical panic grope, and it's like a cross
between a back-seat car game and the SATs.

The only grading marker, though, is self compared to self. If I
were a Keeps participant, I would be on a four-year regimen of some
mystery medication—either estrogen, in one of two forms (estradiol
patches or equine-estrogen pills, to see whether differences emerge
between the two), or placebo patches or placebo pills. Then in another year, I would retake the cognition test, which lasted about an
hour and a half, so researchers could track any change. Brain func-

tion is a major element of the Keeps agenda; the other is heart health, so the test administrators would conduct annual ultrasounds of my carotid artery, to check for the thickening that signals heart disease. That's how they are trying to circumvent the doesn't-manifest-until-you're-older problem, by measuring for known warning markers rather than waiting for the actual big events. They would check my blood and cholesterol for signs of other cardiovascular trouble.

With about 730 participants, Keeps is relatively small; hormone research has been tough to finance in the post-W.H.I. years, and every scientist and physician I've spoken to said there will never again be another hormone trial as costly and ambitious as the W.H.I. A second study, based in Los Angeles, called the Early Versus Late Intervention Trial With Estradiol, is following more than 600 women—comparing a group that has been post-menopausal for an average of 15 years and that is on estradiol or on a placebo with a second, younger group that is an average of three years post-menopausal. "This is the age when we should really study estrogen," says Sanjay Asthana, a University of Wisconsin medical professor who is a designer of the cognition component of Keeps. "People like me are really waiting to see what this data looks like. Either way. We need to know."

Asthana is a geriatrician, with a specialty in Alzheimer's and other forms of age-related memory loss. That makes him a member of what I came to think of, in my travels among estrogen researchers this winter, as the brain contingent. Their working material includes neuroimaging; magnified slices of rodent brains; and live cells that carry on in petri dishes, shooting mitochondria around or struggling under the burden of disease. All these things allow the brain contingent to see, sometimes literally, estrogen in action. It's an amazing process. When cells are healthy, estrogenic molecules slide right in, searching for special receptors that are shaped precisely for the estrogens: the receptors are tiny locks, waiting for the right molecular keys to turn them on. Then, once they are activated by the key-turning process, the work estrogen receptors do is richly complex, if only partly understood. They prod genes into action; they raise good cholesterol; they affect the neurotransmitter chemicals associated with mood and stress, like serotonin and dopamine.

And the brain, scientists have learned in recent decades, is loaded with these receptors. Knowing this makes it easier to understand how perimenopause could start inside aging ovaries and set off such a wild cascade of effects. If you're a typical woman moving through your 40s or 50s, your lifetime egg supply is running out; as that happens, the intricate, multihormone reproductive-signaling loop grows confounded, its triggers altered by the biology of change. The brain and ovaries, the primary stops along this loop, start misreading each other's demands for action. This can make estrogen production crank up frantically, crash and then crank up again. Something also goes awry with most women's thermoregulatory systems, producing hot flashes in around three-quarters of us—nobody yet knows why, exactly, nor why certain women go on flashing for many years while some escape the whole must-remove-outer-garments-now phenomenon entirely. There's an admirably clear explanation of the complete process in a recent book called *Hot Flashes, Hormones and Your Health,* by JoAnn Manson, a Harvard medical professor who worked with both W.H.I. and Keeps. My favorite illustration in Manson's book shows an actual woman's hormone fluctuations as measured before, during and after perimenopause; the "before" graph is a row of calm, evenly spaced ups and downs, various hormones rising and falling in counterpoint and on cue. The lines in the "after" graph are virtually flat. The "during" graph looks as if somebody dynamited a mountain range.

Not all women, Manson notes, experience disruptions as robust as this unidentified patient's. But consider the mess of internal rearrangement we're looking at: the body's overall estrogen production is waning as the ovaries start atrophying into full retirement; and here simultaneously, at least for some of us, is this great Upheaval of

During. The combination of the two can be—how could it not, I thought, the first time I studied the three graphs—a hellacious strain on the brain. Tracing the exact mechanics is still a work in progress, but they surely include some disruption of signaling to the neurotransmitters that make us remember things, experience emotions and generally choreograph the whole thinking operation of the human self.

"There are all these fundamental cognitive functions that many perimenopausal women complain about, and one of those fundamentals is attention," Roberta Brinton, the U.S.C. scientist, told me. "When you can't hold your attention to a thought. Where you're in constant start mode, and you never reach the finish mode. That is devastating."

This was Brinton, as it happens, describing herself. It's why she first went on estrogen (estradiol, accompanied by natural progesterone) when her own perimenopause kicked in a few years ago. We were sitting in a campus garage in her Prius one day, and I asked her what made her so sure her own midlife difficulties—she had the hot flashes, which were obvious, but also the sleep disruption and the infuriating distractibility—were the product of hormonal events, not some womanly existential crisis. We get a lot of that, societally. It's meant to be empathetic. Your role in life is changing, Mrs. Brain Seized by Aliens! Your children are growing up, you're buying expensive wrinkle cream, ice cream makes you gain weight now, of course you're distraught! "Because with estrogen"—Brinton looked at me sharply, and then smiled—"I don't have attention-deficit disorder."

We walked back up to her laboratories, which are spread along a many-roomed warren full of cell incubators, centrifuges and computers. Brinton has thick black hair and a demeanor of lively, good-humored authority; it's easy to envision her as the passionate science professor in crowded lecture halls. But in her labs the work is all rats and mice, many of them surgically or genetically altered to serve as surrogates for adult humans in various stages of maturation or disease. Removing the ovaries from female rats, for example, sends them into low-estrogen mode. Mice can be ordered bred with Alzheimer's. The plaque that clogs the brains of Alzheimer's sufferers, a noxious memory-disrupting substance called beta amyloid, is available as a chemical distillate, which means Brinton's team can experiment with that too—beta amyloid dropped into the brain cells of healthy low-estrogen rodents; or estrogen dropped into cells already damaged by beta amyloid.

That's why Brinton says that the timing hypothesis—the proposition that estrogen could bring great benefit to a woman who starts it in her 50s while having the reverse effect on a woman 10 years older—makes sense even though it is still experimental. She and other scientists know there are ways estrogen improves and protects the brain when it is added to healthy tissue. It makes new cells grow. It increases what's called "plasticity," the brain's ability to change and respond to stimulation. It builds up the density and number of dendritic spines, the barbs that stick out along the long tails of brain cells, like thorns on a blackberry stem, and hook up with other neurons to transmit information back and forth. (The thinning of those spines is a classic sign of Alzheimer's.)

But when estrogen hits cells that are already sick—because they're dying off as part of the natural aging process or because they've been damaged by beta amyloid—something else seems to happen. Dropped in as a new agent, like the wrong kind of chemical solvent sloshed onto rusting metal, estrogen doesn't strengthen or repair. It appears useless. Sometimes it sets off discernible harm. You may recall additional W.H.I. news a few years ago about hormones increasing the risk for aging-related dementia; those stories emerged from a subgroup of W.H.I. participants who were all at least 65 when they started the hormones. There are arguments about that data, like nearly everything else connected to the W.H.I., but the age factor alone reinforces what Brinton and other timing-hypothesis researchers observe in the labs when they give estrogen to ailing cells. "It's like the estrogen is egging on the negative now, rather than the positive," she said. "We know that if you give neurons estrogen, and then expose them to beta amyloid, many more will survive. But when we expose them to amyloid and then give them estrogen—now you don't have survival of the neurons. In some instances, you can actually exacerbate their death."

The heart contingent exploring the timing hypothesis is reasoning the same way. Monkeys get both cardiovascular disease and their own version of menopause; there is a primate team at Wake Forest University in North Carolina that has found estrogen to be a strong protectant for females against future heart disease—but only when it's given at monkey perimenopause. Give estrogen the equivalent of six human years later, says Tom Clarkson, the pathology professor who has been leading this work for decades, and there is no protective effect at all.

Clarkson, who is 78, told me that if he were 30 years younger and a woman, with hot flashes or sleep trouble or sudden crashes of mood, he would have no hesitation about taking hormones. "I absolutely believe in the timing hypothesis," he said. Then, being a scientist, he corrected himself. "I would have to say my level of certainty is 95 percent or greater," he said. "I live a life of believing in the experimental evidence."

So noted, I replied. And what if the symptoms were annoying but bearable or there were no symptoms at all? I've asked the same questions to every researcher I talked to this spring, and nearly all of them reply the same way: if they were deciding for themselves personally, they would tip the risk-benefit scale strongly in favor of hormones as a remedy for immediate ailments of perimenopause. But estrogen solely as a protectant for the heart and brain, to be taken for many years, absent any immediate serious complaints? There was a pause, and I heard Clarkson sigh. "We just don't know about that yet," he said.

* * *

THE PERSONAL CALCULUS OF RISK is an exhausting exercise in the modern era, what with litigation-jumpy physicians, the researchers' candid "We just don't know" and the bottomless learn-it-yourself maw of the Internet. Of all the conversations I had this winter, as I weighed and reweighed the stopping of the patch, the one that most resonates took place on a snowy morning in Washington, in the office of a nursery-school director named Julia Berry. Berry lives not far from the headquarters of the National Institutes of Health in Bethesda, Md., which is why last September she pulled from her mailbox a card the N.I.H. has been mailing to local women within a certain age range. "If you struggle with irritability, anxiety, sadness or loss of enjoyment at the time of the menopausal transition," the card reads, "please call us and help yourself while helping others."

The N.I.H., it turns out, has been quietly conducting mood and hormone studies for more than two decades under the direction of a psychiatrist named Peter Schmidt and his predecessor, David Rubinow, who is now chairman of the psychiatry department at the University of North Carolina. The research was first set into motion by Rubinow's postgraduate interest in premenstrual syndrome; the idea of giving younger women drugs to lower and flatten temporarily their estrogen and progesterone levels, essentially inducing menopause, was initially conceived to determine the role of hormones in PMS—to see whether these young women got relief when their hormones stopped the cresting and dropping of the normal menstrual cycle. (It often worked as a short-term treatment and yes, the young women often got hot flashes.) In recent years, the induced-menopause experiments have continued, among many other studies, as part of an effort to try to understand the chemistry of women like Julia Berry and me—women for whom perimenopause turns into what Berry described to me as "psychological misery, not myself and absent from the world."

Berry is 55, ponytailed and roundish and pretty. She was divorced a long time ago, raised three good kids mostly on her own and has a firm handshake and a job she loves. Her troubles started in her late 40s, in the standard way, with hot flashes and jerking awake at 3 a.m. and then escalated into something much fiercer. Like me, at the worst of it, she occasionally found herself in traffic, wishing silently for an oncoming truck that might exit her swiftly from this life without qualifying as a suicide. A physician prescribed antidepressants. They helped, with both the anguish and the flashes, but not enough. "I am one of the most steady, even-keeled, hard to ruffle, really unflappable . . . truly," Berry told me. "I am. I, generally speaking, can be completely relied upon to do the sensible right thing almost all the time. Which is one of the reasons this period in my life has been so weird."

She called the N.I.H. number at once. She was quickly evaluated, enrolled in a double-blind study of the effects of estrogen on perimenopausal depression and sent home with a paper bag containing a mystery patch. When I asked Berry to describe the sensation of the next few weeks, she looked up at the ceiling for a second to think. "Kind of like having been in a smoky room, waving your arms and now seeing that the exhaust fan is taking a little at a time," she said. "My mood lifted. First time in three years I wasn't waking up at 3 in the morning. That's when I knew I wasn't on the placebo. It was very clear to me that there was something fundamentally wrong with my chemical systems, and that whatever was in this patch was setting things right, so that I could function like a regular human being— the human being I was familiar with."

What medicine doesn't know about the chemistry of mood, including clinical depression, dwarfs what medicine doesn't know about hormones. It would be handy for science if Berry and I could have made our heads available for dissection at certain points in recent years; as it is, we're able to answer as many elaborations on "I feel bad" or "I feel good" as researchers might wish to throw at us, but they still have no way of pinning down where we belong on the scale of menopausal distress, or what exactly we're doing there. We could be extra-high-volume versions of the women who are having an ordinary rough time of it, like Roberta Brinton—the women who hot-flash and can't sleep and cast about for vocabulary with which to describe feeling, as Brinton puts it, "just off." Or we could belong to some subcategory of anomalies, women with a wired-in susceptibility to depression—gene pools, childhoods, whatever—that was fired up by abrupt hormonal change.

Some psychological surveys will tell you there's no evidence for a surge of clinical depression at menopause. I believe that, given how many other phases of life can unhinge us, but I also believe—no, actually, I know—that there is a difficult thing that happens to some women in the perimenopausally affected brain. Hostile as I am to generalizations involving women rendered fragile by biology, here I am, and here, too, is Berry, both of us pulled out of something terrible by a pharmaceutical infusion of estrogen. Two physicians who specialize in hormones and mood, Louann Brizendine, a neuropsychiatrist at the University of California, San Francisco, and Claudio Soares, a Canadian research psychiatrist who works at McMaster University in Ontario, told me that women who seek them out tell variations of the same story Berry and I took to our doctors: I know that something is wrong with me because I also know, somewhere in the noncrazy part of myself, that there is such pleasure to be offered by the circumstances of my grown-up life.

"These women thought they were losing their minds," Brizendine told me, describing the 40-to-60-year-old patients she began seeing when she opened the Women's Mood and Hormone Clinic at the university in 1994. "In 1994 we didn't have words for it," she said. "Now we do. It's called perimenopausal depression."

Brizendine and Soares, like Schmidt and Rubinow, have found that various combinations work with varying degrees of effectiveness for many of us—hormones with an antidepressant, hormones without an antidepressant, sometimes antidepressants on their own. The alternatives-to-hormones recommendations are mostly fine things in their own right, varying from certainly useful to harmless: exercise regularly, keep the weight down, easy on the caffeine, calm yourself with deep breathing or yoga, try black cohosh. (You could start a bar brawl over the efficacy of black cohosh, but the general consensus seems to be: if it works for you, go for it.) But the troubles set off by ricocheting hormones are reliably fixed by making the hormones stop ricocheting. And the laborious weighing of hormones' benefits versus hormones' harms—maybe not at the crisis moment, for those of us at our most distraught, but later, one or two or five years down the road—is something still undertaken by millions of women along the full breadth of the perimenopausal spectrum.

HOW IN THE WORLD TO do it wisely enough so the calculation is as right for each of us as it can possibly be? JoAnn Manson's book contains the most careful checklist I've seen yet; by the time you answer all the personal-history questions the book asks you to consider, you've read 82 pages. Breast cancer is a factor, to be sure, but so are colorectal cancer, ovarian cancer, stroke, hip fracture and diabetes. If the timing hypothesis proves right and estrogen really does protect our brains and our hearts as long as we start it early enough, the calculation only grows that much more important and complex. There are moving pieces involved in working out every one of these risks in relation to everything else, and anyone who thinks there's a bumper-sticker answer to the hormones question—don't take them, you're sure to be better off—is, like me that day in the psych unit, neither listening to scientific argument nor reading the fine print.

Here's one example from the many to which researchers have pointed me this winter. Remember MPA? The synthetic progesterone-like substance used along with equine estrogens in the W.H.I.? There was a second W.H.I. hormones-versus-placebo trial, of nearly 11,000 women, that was also started in the early 1990s, just like the one that was halted in 2002. All the women enrolled in this second study had undergone hysterectomies, which meant they had zero risk for uterine cancer. So the women on medications in this trial were taking only equine estrogens—no MPA, which you'll recall is given to protect the uterus. Their study was stopped in 2004, also before its planned end date, because the estrogen-taking women were showing a higher risk of stroke than the women on the placebo. But their breast-cancer rate was lower. The hormone-taking women with hysterectomies in that second study, who used estrogen without MPA, showed a 23 percent lower risk of invasive breast cancer than their counterparts who were taking no hormones at all.

Nobody's persuaded that this means MPA promotes breast cancer while estrogen does not. It's clear that estrogen acts aggressively on certain breast malignancies and that any woman who has had breast cancer or has a history of it in her immediate family should stay off estrogen. This is one of the principal reasons such intense work is under way right now, in labs like Roberta Brinton's, to develop estrogenic variants—molecular substances designed to latch only to certain receptors (in the brain, say, where the activated receptors can do their good works) while ignoring receptors in the breast and uterus. And there are plenty of confounding factors, as scientists say, with regard to the women in the no-MPA trial. They all had undergone hysterectomies, for one thing; maybe whatever caused them to require uterine removal in the first place affected their reactions to the estrogen.

Or it could have been a fluke. But the MPA wrinkle adds suspicion and urgency to the timing-hypothesis questions about what really goes on when women of our demographic use hormones, and Julia Berry and I spent a long time talking about this, the adding and subtracting, the guessing and weighing, the balancing of what we think we know about ourselves against what we cannot possibly foresee. We will both, for the present, continue wearing estrogen patches. Berry turned out to be right, of course; she wasn't on the placebo, which the N.I.H. doctors told her when she finished the study. And as she hurried to fill her own patch prescription, she found her gratitude mixed with more than a little frustration. "Why did my primary-care physician give me an antidepressant when I could have had something simple, like estrogen?" she asked. "Why don't they know?"

We talked about breast cancer, because that is the nightmare illness in nearly all our calculations, for most of us the visual closest to hand. Three of my best friends have endured the full breast-cancer horror show and by now have retired their wigs. All have survived. None had been on hormone replacement. This is information that batters me steadily but not helpfully, like my ex-smoker paternal aunt's fatal lung cancer and the fact that I'm a lifetime nonsmoker and regular exerciser with extremely good cholesterol levels. How do my lowered risks from one column balance against my question marks over in another column? What to do?

"I'd rather monitor something I know can go wrong than go on living in the state I was in," Berry said. "I could have my breasts removed. I like them. But they're not my life."

We've spent a fair bit of time by now, Julia Berry and I, shaking these uncertainties out and squinting at them. Do we wear these patches forever? We don't know. What happens when we do take them off, if we do? We don't know. Have we done nothing except delay a biological process, complete with hot flashes and another round of truck-crash fantasies, that at some point we'll have to bully our way through? We don't know, nor does any researcher I talked to this spring.

And there's this: Should luck and longevity cooperate, we are going to grow old. We're already old, by the standards of our children and our ancestors, but the generation to which we belong expects to live a rich messy life full of extremely loud rock music for another 30 years after menopause. Every midlife woman I know keeps redrawing for herself the defensible lines of intervention in the "natural" sequence of human aging. Obsessive multiple plastic surgeries are silly and desperate. Muscles kept in good working order are not. Where on that spectrum is a hormones-saturated pharmaceutical patch? What if the timing hypothesis is even partly right? Suppose all we learn about replacement estrogen, in the end, is that if it's started early enough it might protect the heart and the brain, and that its chemistry makes some of us feel more the way we did at 40 than the way our mothers did at 65? Not an elixir of youth. More like . . . reading glasses. Or calcium supplements, or painkillers that stop the knee from hurting but carry risk warnings of their own. It has occurred to me that the better analogy might be a 13-year-old trying to ward off puberty by binding her breasts, but most of the time I don't think so, and if I do try stopping the patches, I know this to a certainty: I will keep a few extras in reserve, just in case.

CHARLES SIEBERT

The Animal-Cruelty Syndrome

FROM *THE NEW YORK TIMES MAGAZINE*

Law-enforcement agencies and researchers are beginning to see clear connections between animal abuse and violent crime. Charles Siebert investigates what might be behind this link.

ON A LATE MAY AFTERNOON LAST YEAR IN southwest Baltimore, a 2-year-old female pit bull terrier was doused in gasoline and set alight. A young city policewoman on her regular patrol of the neighborhood of boarded-up row houses and redbrick housing developments turned her squad car onto the 1600 block of Presbury Street and saw a cloud of black smoke rising from the burning dog. She hopped out, ran past idle onlookers and managed to put out the flames with her sweater. The dog, subsequently named Phoenix, survived for four days with burns over 95 percent of her body, but soon began to succumb to kidney failure and had to be euthanized.

It was only a matter of hours before the story, made vivid by harrowing video footage of the wounded dog, was disseminated nationwide in newspapers, TV and radio newscasts and countless Web sites. An initial $1,000 reward for the capture of the culprits would soon climb to $26,000 as people around the country followed Phoenix's struggle for life. A gathering of people in Venice Beach, Calif., held a candlelight vigil for her. A month later, the mayor of Baltimore, Sheila Dixon, announced the creation of the Anti-Animal-Abuse Task Force to work in concert with city officials, local law enforcement and animal rights and animal-control groups to find ways to better prevent, investigate and prosecute such crimes.

The scale, speed and intensity of the response were striking. The subject of animal abuse, especially the abuse of pit bulls in dog-fighting activities, has achieved a higher profile after the 2007 arrest of the N.F.L. star Michael Vick for operating an illegal interstate dog-fighting operation in Surry County, Va. But the beleaguered pit bull is merely the most publicized victim of a phenomenon that a growing number of professionals—including police officers, prosecutors, psychologists, social workers, animal-control officers, veterinarians and dogcatchers—are now addressing with a newfound vigor: wanton cruelty toward animals. Before 1990, only six states had felony provisions in their animal-cruelty laws; now 46 do. Two years ago, the American Society for the Prevention of Cruelty to Animals formed the nation's first Mobile Animal Crime Scene Investigation Unit, a rolling veterinary hospital and forensic lab that travels around the country helping traditional law-enforcement agencies follow the evidentiary trails of wounded or dead animals back to their abusers.

In addition to a growing sensitivity to the rights of animals, another significant reason for the increased attention to animal cruelty is a mounting body of evidence about the link between such acts and serious crimes of more narrowly human concern, including illegal firearms possession, drug trafficking, gambling, spousal and child abuse, rape and homicide. In the world of law enforcement—and in the larger world that our laws were designed to shape—animal-cruelty issues were long considered a peripheral concern and the province of local A.S.P.C.A. and Humane Society organizations; offenses as removed and distinct from the work of enforcing the human penal code as we humans have deemed ourselves to be from animals. But that illusory distinction is rapidly fading.

"With traditional law enforcement," Sgt. David Hunt, a dog-fighting expert with the Franklin County Sheriff's Office in Columbus, Ohio, told me, "the attitude has been that we have enough stuff on our plate, let the others worry about Fluffy and Muffy. But I'm starting to see a shift in that mentality now." Hunt has traveled to 24 states around the country in order to teach law-enforcement personnel about the dog-fighting underworld, often stressing the link between activities like dog fighting and domestic violence. "You have to sell it to them in such a way that it's not a Fluffy-Muffy issue," he said of teaching police officers about animal-abuse issues. "It's part of a larger nexus of crimes and the psyche behind them."

The connection between animal abuse and other criminal behaviors was recognized, of course, long before the evolution of the social sciences and institutions with which we now address such behaviors. In his famous series of 1751 engravings, "The Four Stages of Cruelty," William Hogarth traced the life path of the fictional Tom Nero: Stage 1 depicts Tom as a boy, torturing a dog; Stage 4 shows Tom's body, fresh from the gallows where he was hanged for murder, being dissected in an anatomical theater. And animal cruelty has long been recognized as a signature pathology of the most serious violent offenders. As a boy, Jeffrey Dahmer impaled the heads of cats and dogs on sticks; Theodore Bundy, implicated in the murders of some three dozen people, told of watching his grandfather torture animals; David Berkowitz, the "Son of Sam," poisoned his mother's parakeet.

But the intuitions that informed the narrative arc of Tom Nero are now being borne out by empirical research. A paper published in a psychiatry journal in 2004, "A Study of Firesetting and Animal Cruelty in Children: Family Influences and Adolescent Outcomes," found that over a 10-year period, 6-to-12-year-old children who were described as being cruel to animals were more than twice as likely as other children in the study to be reported to juvenile authorities for a violent offense. In an October 2005 paper published in *Journal of Community Health*, a team of researchers conducting a study over seven years in 11 metropolitan areas determined that pet abuse was one of five factors that predicted who would begin other abusive behaviors. In a 1995 study, nearly a third of pet-owning victims of domestic abuse, meanwhile, reported that one or more of their children had killed or harmed a pet.

The link between animal abuse and interpersonal violence is becoming so well established that many U.S. communities now cross-train social-service and animal-control agencies in how to recognize signs of animal abuse as possible indicators of other abusive behaviors. In Illinois and several other states, new laws mandate that veterinarians notify the police if their suspicions are aroused by the condition of the animals they treat. The state of California recently added Humane Society and animal-control officers to the list of professionals bound by law to report suspected child abuse and is now considering a bill in the State Legislature that would list animal abusers on the same type of online registry as sex offenders and arsonists.

When I spoke recently with Stacy Wolf, vice president and chief legal counsel of the A.S.P.C.A.'s Humane Law Enforcement department, which focuses on the criminal investigation of animal-cruelty cases in New York City, she drew a comparison between the emerging mindfulness about animal cruelty and the changing attitudes toward domestic abuse in the 1980s. "It really has only been in recent years that there's been more free and accurate reporting with respect to animal cruelty, just like 30 years ago domestic violence was not something that was commonly reported," she said. "Clearly every act of violence committed against an animal is not a sign that somebody is going to hurt a person. But when there's a pattern of abusive behavior in a family scenario, then everyone from animal-control to family advocates to the court system needs to consider all vulnerable victims, including animals, and understand that violence is violence."

It isn't clear whether Phoenix was used for dog fighting. Subsequent examinations of her body did find—along with evidence that gasoline had been poured down her throat—a number of bite wounds. Veterinarians, however, said that those could have been self-inflicted in the course of Phoenix's frenzied attempts to fight off the flames. But prosecutors also later claimed that Phoenix's accused assailants, 17-year-old twin brothers named Tremayne and Travers Johnson, of a nearby block of Pulaski Street, were using a vacant neighborhood home for the keeping of pit bulls and other ganglike activities.

The Johnson twins have pleaded not guilty. According to court documents, both suspects, said to be members of the 1600 Boys gang, were identified by a witness as running out of the alley where the dog was set alight. "There was some gang-style graffiti found in that abandoned building," Randall Lockwood, the A.S.P.C.A.'s senior vice president for forensic sciences and anticruelty projects, and a member of the new Anti-Animal-Abuse Task Force in Baltimore, told me at the A.S.P.C.A.'s Midtown Manhattan offices in December. "There was also dog feces on the premises. Unfortunately, nobody bothered collecting the feces to see if it was from Phoenix."

Along with the need to track the physical evidence of animal cruelty there is the deeper and more complex challenge of trying to parse its underlying causes and ultimate ramifications. As a graduate student in psychology, Lockwood had an interest in human-animal interactions and the role of animals and education in the development of empathy in children. This inevitably led him to consider the flip side of the equation: the origins of cruelty to animals and what such behavior might indicate about an individual's capacity for empathy and his or her possible future behavior.

Back in the early 1980s, Lockwood was asked to work on behalf of New Jersey's Division of Youth and Family Services with a team of investigators looking into the treatment of animals in middle-class American households that had been identified as having issues of child abuse. They interviewed all the members of each family as well as the social workers who were assigned to them. The researchers' expectation going in was that such families would have relatively few pets given their unstable and volatile environments. They found, however, not only that these families owned far more pets than other households in the same community but also that few of the animals were older than 2.

"There was a very high turnover of pets in these families," Lockwood told me. "Pets dying or being discarded or running away. We discovered that in homes where there was domestic violence or physical abuse of children, the incidence of animal cruelty was close to 90 percent. The most common pattern was that the abusive parent had used animal cruelty as a way of controlling the behaviors of others in the home. I've spent a lot of time looking at what links things like animal cruelty and child abuse and domestic violence. And one of the things is the need for power and control. Animal abuse is basically a power-and-control crime."

The dynamic of animal abuse in the context of domestic violence is a particularly insidious one. As a pet becomes an increasingly vital member of the family, the threat of violence to that pet becomes a strikingly powerful intimidating force for the abuser: an effective way for a petty potentate to keep the subjects of his perceived realm in his thrall. In 2005, Lockwood wrote a paper, "Cruelty Toward

Cats: Changing Perspectives," which underscores this dynamic of animal cruelty as a means to overcome powerlessness and gain control over others. Cats, Lockwood found, are more commonly victims of abuse than dogs because dogs are, by their very nature, more obedient and eager to please, whereas cats are nearly impossible to control. "You can get a dog to obey you even if you're not particularly nice to it," Lockwood told me. "With a cat you can be very nice, and it's probably going to ignore you, and if you're mean to it, it may retaliate."

Whatever the particular intimidation tactics used, their effectiveness is indisputable. In an often-cited 1997 survey of 48 of the largest shelters in the United States for victims of domestic violence and child abuse, more than 85 percent of the shelters said that women who came in reported incidents of animal abuse; 63 percent of the shelters said that children who came in reported the same. In a separate study, a quarter of battered women reported that they had delayed leaving abusive relationships for the shelter out of fear for the well-being of the family pet. In response, a number of shelters across the country have developed "safe haven" programs that offer refuges for abused pets as well as people, in order that both can be freed from the cycle of intimidation and violence.

What cannot be so easily monitored or ameliorated, however, is the corrosive effect that witnessing such acts has on children and their development. More than 70 percent of U.S. households with young children have pets. In a study from the 1980s, 7-to-10-year-old children named on average two pets when listing the 10 most important individuals in their lives. When asked "whom do you turn to when you are feeling sad, angry, happy or wanting to share a secret," nearly half of 5-year-old children in another study mentioned their pets. One way to think of what animal abuse does to a child might simply be to consider all the positive associations and life lessons that come from a child's closeness to a pet—right down to eventually receiving their first and perhaps most gentle experiences of death as a natural part of life—and then flipping them so that all those lessons and associations turn negative.

In a 2000 article for *AV* Magazine, a publication of the American Anti-Vivisection Society, titled, "Wounded Hearts: Animal Abuse and Child Abuse," Lockwood recounts an interview he conducted for the New Jersey Division of Youth and Family Services in the early 1980s. He describes showing to "a perky 7-year-old boy" a simple drawing of a boy and a dog, playing ball inside a house and a broken lamp on the floor beside them. Lockwood asked the 7-year-old—a child who had witnessed his brother being beaten by their father, who was "reportedly responsible for the 'disappearance' of several family pets"—to describe what would happen next in the story of the boy in the picture. "He grew still and sullen," Lockwood writes, "and shook his head slowly. 'That's it,' he said in a matter-of-fact tone, 'They're all going to die.'"

Children who have witnessed such abuse or been victimized themselves frequently engage in what are known as "abuse reactive" behaviors, Lockwood said, re-enacting what has been done to them either with younger siblings or with pets. Such children are also often driven to suppress their own feelings of kindness and tenderness toward a pet because they can't bear the pain caused by their own empathy for the abused animal. In an even further perversion of an individual's healthy empathic development, children who witness the family pet being abused have been known to kill the pet themselves in order to at least have some control over what they see as the animal's inevitable fate. Those caught in such a vicious abuse-reactive cycle will not only continue to expose the animals they love to suffering merely to prove that they themselves can no longer be hurt, but they are also given to testing the boundaries of their own desensitization through various acts of self-mutilation. In short, such children can only achieve a sense of safety and empowerment by inflicting pain and suffering on themselves and others.

*   *   *

IN MARCH I PAID A visit to the newly established Veterinary Forensics Medicine Sciences program at the University of Florida,
Gainesville. Directed by Melinda Merck, a veterinarian who serves as
the A.S.P.C.A.'s senior director of veterinary forensics and as the
"captain" of its new mobile C.S.I. unit, the program is the first of its
kind at a major U.S. university. As animal abuse has become an increasingly recognized fixture in the context of other crimes and their
prosecution, it is also starting to require the same kinds of sophisticated investigative techniques brought to bear on those other crimes.

Veterinary forensic students at the University of Florida are being
trained in the same way that traditional crime-scene investigators
are, taking courses in a wide range of topics: crime-scene processing;
forensic entomology (determining the time of an animal's injury or
death by the types of insects around them); bloodstain-pattern and
bite-mark analysis; buried-remains excavation; and forensic osteology (the study of bones and bone fragments).

"I love being around bones," Merck proclaimed as she led me into
the university's C. A. Pound Human Identification Laboratory, a
sprawling, brashly lighted array of human skeletal remains arranged
in meticulous piecemeal patterns on rows of shiny metal tables. "I
find bones fascinating. There is a lot of information in them." Merck,
who testifies at animal-cruelty trials across the country, conducted
the forensic osteology on the dog remains recovered from the mass
graves on Michael Vick's Virginia property in 2007.

The lab is one of the busiest of its kind in the world, enlisted for
countless crime-scene investigations and archaeological digs and to
help identify the victims of disasters, including those of the 9/11 attacks on the World Trade Center and Hurricane Katrina. The fact
that one of the examining tables and adjacent bone-boiling and
cleansing units have now been assigned to Merck for her own ani-
mal-forensic work and course instruction speaks volumes about the
shifting perspective toward animal-cruelty crimes. "We have a really

cool thing going on here," Merck told me. "We have the collaborative effort of a lot of big-wig forensic specialists down here with years of experience."

She led me over to her examining table. Set at one end was what she called "my box of evidence," a picnic-cooler-size plastic container that held the excavated remains from a mass grave, part of an investigation she is conducting into a suspected dog-fighting operation in Georgia. "In most of our cases of animal cruelty, the bodies are not fresh," she said. "They're decomposed. They're discarded. They're hidden. And so the advanced post-mortem stage is where we really need to be experts."

Merck's 2006 book, *Forensic Investigation of Animal Cruelty: A Guide for Veterinary and Law Enforcement Professionals,* which she wrote with Randall Lockwood and Leslie Sinclair of Shelter Veterinary Services in Columbia, Md., contains a daunting list of the grisly things human beings do to animals: thermal injuries (immolation, baking, microwaving); blunt-force trauma; sharp-force and projectile injuries; asphyxiation; drowning; poisoning; ritual murders; and sexual assault. Merck spared no details in discussing such horrors over the course of a veterinary-forensics lecture I attended earlier that day, held in a conference room at a hotel near the university as part of a four-day seminar. Even Merck's seasoned audience of out-of-town vets, A.S.P.C.A. disaster-response and investigative-team workers, community-outreach personnel and the chief legal counsel for New York City's Humane Law Enforcement department could be heard gasping into their coffee mugs as Merck annotated, one after the next, screen-projected slides of stark brutality: blood-drenched dog-fighting pits; bludgeoned, internally hemorrhaging pets; bruised and mutilated canine sexual organs; a heavily duct-taped, paint-coated puppy and the fur-lined, nail-scraped oven walls from which the puppy struggled vainly to escape.

Those whose compassion compels them to confront and combat daily its utter absence are, of necessity, often forced to affect a passionless pose. Merck proceeded through her seminar with clinical speed and precision through a series of signature forensic cases. One of the first pivoted around the mystery of a missing Pomeranian whose owners were convinced had been stolen from their backyard. Merck called up the slide of a tiny skeleton she had rendered in her corner of the lab from remains found in a vacant lot not far from the Pomeranian owners' home. It looked like a wingless bat, the delicate brace of ribs bearing tiny symmetrical snaps on each side.

"What could have caused these," Merck asked, pointing her red laser at the breaks. "What could make a dog disappear so fast?"

"Man!" someone called out to bursts of laughter.

"What else," Merck said, smiling.

"A bird of prey!"

"Yep," Merck nodded. "Most likely a hawk. These two breaks are where the bird's talons grabbed hold of the dog. This is why forensic osteology is so important, and yet there's nothing in our standard veterinary training that teaches us how to look at bones properly."

Merck soon proceeded to the case of the puppy found four years ago in the oven of a ransacked community center in Atlanta. An outraged local prosecutor called Merck about the case and then showed up at her vet clinic one day with the dog's remains. "She brings me the puppy, and this . . . ," Merck said, the slide behind her now sapping the room's air, "is what she brings me."

Step by step, from the outer paint to the unraveled layers of duct tape to the dog's abraded nails and paws to the hem of an old T-shirt that was used as a leash, Merck's detailed forensic analysis of the victim and of the crime scene would be used to assemble a timeline of events. Ultimately, her analysis would help seal the conviction of two teenage brothers on multiple charges, including burglary, animal cruelty and—because the brothers had shown a number of children at the community center what they had done and then threatened them with their lives if they told anyone—additional charges of child abuse and terroristic threats.

* * *

THE MOST COMMON DYNAMIC BEHIND the cases cited that morning was that of a man abusing a family pet to gain control over, or exact revenge against, other family members. Merck told of one puppy found buried in the backyard of a house. As Merck tells it, the dog belonged to the female friend of a woman who had recently left the man with whom she and her two children from a previous marriage were living. She and her children had moved in with the friend, someone who the man decided was keeping him and his estranged partner from reuniting. The girlfriend's pet, therefore, became for him the optimum vehicle for expressing his rage against both women.

"He tortured the puppy when the two women weren't home," Merck told me after her lecture that day. "He also tried to make two of the kids participate just to make it more heinous. So along with the animal cruelty, of course, we had child abuse."

Merck has made it her mission to urge other vets to report and investigate suspected cases of animal abuse, incorporating a few cautionary tales of her own into her lectures to point up the often dire consequences of failing to do so. One involved a man from Hillsborough County in Florida who was arrested for murdering his girlfriend, her daughter and son and their German shepherd. He had previously been arrested (but not convicted) for killing cats. In another story Merck tells, one related to her by a New York City prosecutor, a woman reported coming home to find her boyfriend sexually molesting her Labrador retriever, but the case never went to trial.

"My point on that one," Merck told me, "is that no one took precautions to preserve the evidence on the dog. And once it comes down to a he-said-she-said type of situation, you're lost. These types of cases are difficult enough even when we have all the evidence, in part because it's very hard for investigators and prosecutors to even consider that someone would do things like this. It's so disturbing and offensive, they don't know what to do about it. A lot of the work

I do involves not just talking to vets but reaching out to law enforcement to make them more knowledgeable on these matters, to make them understand, for example, that things like sexual assault of children and animals are linked. They are similar victims."

On our way back to the hotel for an afternoon lecture on forensic entomology, Merck made a little detour to show me the A.S.P.C.A.'s new mobile C.S.I. unit, parked in a side lot of the vet school's farm-animal compound. Twenty-six-feet long, with its own climate-control, generator, examination room and surgical suite, digital microscope, X-ray machine, sexual-assault kit and anesthesia-oxygen machine, it is essentially a giant emergency room on wheels, allowing Merck and her crew to examine and care for animals at suspected crime scenes and to efficiently analyze and process evidence to ensure its integrity.

The van was an important part of the largest dog-fighting raid in American history last year, in which more than 400 dogs were rescued and 26 people from six states arrested. "We had two forensic teams on board for that," Merck said. "We had to hit 25 different crime scenes in one day. We hit the first one at 7 a.m., and we finished up at around 6 a.m. the following morning."

When I asked Merck if she thought incidents of animal cruelty were on the rise or if it was that we are now being more vigilant about them, she said that it is probably more the latter. "We're more aware now," she said, "but there is also more of a support system for responding to these incidents. When I started out as a vet 20 years ago, I was one of the few who would call if I got a suspicious case, and that was when such things were still a misdemeanor and it wasn't law enforcement involved. It was animal control taking care of nuisance animals. Now with veterinarians I tell them you cannot not report, because you don't know if what you're seeing on the animal isn't the proverbial tip of the iceberg."

Merck then recalled for me a personal experience she most likes to relate in classes and seminars, what she's dubbed "the tale of the good

Samaritan and the savvy vet." An Atlanta contractor pulled up to a house one morning where he was to perform some work. As he got out of his truck, he heard a dog screaming from the house next door, went over to investigate and saw through an open garage door a dog dragging its back legs and a woman standing beside it. The woman instantly began pleading to the contractor that the dog needed to be euthanized, but she said she couldn't afford the vet bills. The contractor offered to take the dog to his vet, who, upon examining the dog, agreed that it was too debilitated to be saved. He then told the contractor that there was something suspicious about the case and that he was going to report it to animal services for whom Merck worked at the time as a consultant outside of her daily vet practice.

"THEY ASKED ME TO PERFORM a necropsy," Merck told me. "It turns out the dog was paralyzed from having been beaten so often. I reported what I found. Police went to the woman's house to make an arrest. They found a badly bruised boy. And just like that both parents are being hauled off for child abuse. So there was a classic case of the system working like it should."

Last November, Lockwood was asked to testify at the pretrial hearing in which a judge ruled that Tremayne and Travers Johnson would be tried as adults for the burning of Phoenix in Baltimore last year. Lockwood looked at dozens of pictures of Phoenix in order to select which images to present to A.S.P.C.A. staff members. "I could only find one that wasn't overwhelmingly disturbing," he told me. "It's where she's so bundled up in gauze and bandages you can't really see anything. It's easy to empathize with burns because we've all been burned, and even if it's only minor, you realize how painful that is."

The matter of empathy, of course, goes to the heart of most of our inquiries into the nature of cruel acts and their possible causes. There seems to be little doubt anymore about the notion that a person's capacity for empathy can be eroded; that someone can have, as

Lockwood put it to me, "their empathy beaten or starved out of them." To date, little is known about the Johnson twins' background beyond the fact that they both reportedly have chronic truancy issues and previous probation violations and were recently involved with a gang. Along with possible early abuse or genetic and biological components, Lockwood also spoke of the frequent association between environment and acts of violence, how poverty often creates the sense of persecution and injustice that makes some people feel justified in striking back in order to gain the sense of power and control they otherwise lack.

"What I have the most trouble relating to," Lockwood told me, "and the Phoenix kids might be indicative of this sort of thing, is the kind of cruelty that happens just out of boredom. I've had quite a few cases where I ask a kid, Why did you blow up that frog or set fire to that cat? and they don't respond with answers like 'I hate cats' or 'I didn't see that as a living thing.' Their answer is 'We were bored.' And then you have to ask yourself, Well, what about alternative pathways to alleviating this boredom? I have difficulty grasping what would be the payoff for setting fire to a dog."

NEUROSCIENTISTS ARE NOW BEGINNING TO get a fix on the physical underpinnings of empathy. A research team at the University of Chicago headed by Jean Decety, a neuroscientist who specializes in the mechanisms behind empathy and emotional self-regulation, has performed fMRI scans on 16-to-18-year-old boys with aggressive-conduct disorder and on another group of similarly aged boys who exhibited no unusual signs of aggression.

Each group was shown videos of people enduring both accidental pain, like stubbing a toe, and intentionally inflicted pain, like being punched in the arm. In the scans, both groups displayed a similar activation of their empathic neural circuitry, and in some cases, the boys with conduct disorder exhibited considerably more activity than those in the control group. But what really caught the attention of the researchers was the fact that when viewing the videos of intentionally inflicted pain, the aggressive-disorder teenagers displayed extremely heightened activity in the part of our brain known as the reward center, which is activated when we feel sensations of pleasure. They also displayed, unlike the control group, no activity at all in those neuronal regions involved in moral reasoning and self-regulation.

"We're really just beginning to have an inkling of the neurophysiology of empathy," Lockwood told me. "I think empathy is essentially innate, but I also think empathy can be learned, and I know it can be destroyed. That's why having a better understanding of the neurophysiology will really help us. Just doing a social intervention on a person doesn't do any good if you're not aware of certain physiological deficits. As I heard someone put it at a recent lecture I attended, that would be like an orthopedist telling someone with a broken arm to lift weights. It won't do anything until the arm is set, and it actually might make things worse. I try to understand who the kids are who seem beyond reach, who seem to have truly impaired systems of empathy. And then I ask, Can that be restored?"

It turns out that just as recent brain-imaging studies have begun to reveal the physical evidence of empathy's erosion, they are now also beginning to show definitive signs of its cultivation as well. A group of researchers led by Richard Davidson, a professor of psychiatry and psychology at the University of Wisconsin, Madison, published a study in a March 2008 edition of the *Public Library of Science One*, showing that the mere act of thinking compassionate thoughts caused significant activity and physical changes in the brain's empathic pathways. "People are not just stuck at their respective set points," Davidson has said of the study's results. "We can take advantage of our brain's plasticity and train it to enhance these qualities. . . . I think this can be one of the tools we use to teach emotional regulation to kids who are at an age where they're vulnerable to going seriously off track."

To date, one of the most promising methods for healing those whose empathic pathways have been stunted by things like repeated exposure to animal cruelty is, poetically enough, having such victims work with animals. Kids who tend to be completely unresponsive to human counselors and who generally shun physical and emotional closeness with people often find themselves talking openly to, often crying in front of, a horse—a creature that can often be just as strong-willed and unpredictable as they are and yet in no way judgmental, except, of course, for a natural aversion to loud, aggressive human behaviors.

Equine-therapy programs, for example, are now helping an increasing number of teenagers who have severe emotional and behavioral issues, as well as children with autism and Asperger's syndrome. At Aspen Ranch in Loa, Utah, troubled teenagers are being paired off with wild mustangs that have been adopted from the Bureau of Land Management, each species ultimately managing to temper the other, a dynamic that has also proved very effective in teaching patience and empathy to prisoners in correctional facilities. In the Los Angeles suburb of Compton, there is a youth equestrian program called the Compton Junior Posse. Teenagers clean stables, groom horses and then ride them in amateur equestrian events across Southern California. There are now bovine- and elephant-assisted therapy programs as well.

For Lockwood, animal-therapy programs draw on the same issues of power and control that can give rise to animal cruelty, but elegantly reverse them to more enlightened ends. "When you get an 80-pound kid controlling a 1,000-pound horse," he said, "or a kid teaching a dog to obey you and to do tricks, that's getting a sense of power and control in a positive way. We all have within us the agents of entropy, especially as kids. It's easier to delight in knocking things down and blowing stuff up. Watch kids in a park and you see them throw rocks at birds to get a whole cloud of them to scatter. But to lure animals in and teach them to take food from your hand or to obey commands, that's a slower process. Part of the whole enculturation and socialization process is learning that it's also cool and empowering to build something. To do something constructive."

AMY HARMON

A Soft Spot for Circuitry

FROM *THE NEW YORK TIMES*

As robots become more intelligent, and more responsive to human interaction, they are being enlisted to perform tasks with interpersonal elements—like therapy. Rather than posing a threat, these robots seem to be delighting the people who use them, as Amy Harmon reports.

NOTHING EILEEN OLDAKER TRIED COULD CALM her mother when she called from the nursing home, disoriented and distressed in what was likely the early stages of dementia. So Ms. Oldaker hung up, dialed the nurses' station and begged them to get Paro.

Paro is a robot modeled after a baby harp seal. It trills and paddles when petted, blinks when the lights go up, opens its eyes at loud noises and yelps when handled roughly or held upside down. Two microprocessors under its artificial white fur adjust its behavior based on information from dozens of hidden sensors that monitor sound, light, temperature and touch. It perks up at the sound of its name, praise and, over time, the words it hears frequently.

"Oh, there's my baby," Ms. Oldaker's mother, Millie Lesek, exclaimed that night last winter when a staff member delivered the seal to her. "Here, Paro, come to me."

"Meeaakk," it replied, blinking up at her through long lashes.

Janet Walters, the staff member at Vincentian Home in Pittsburgh who recalled the incident, said she asked Mrs. Lesek if she would watch Paro for a little while.

"I need someone to baby-sit," she told her.

"Don't rush," Mrs. Lesek instructed, stroking Paro's antiseptic coat in a motion that elicited a wriggle of apparent delight. "He can stay the night with me."

After years of effort to coax empathy from circuitry, devices designed to soothe, support and keep us company are venturing out of the laboratory. Paro, its name derived from the first sounds of the words "personal robot," is one of a handful that take forms that are often odd, still primitive and yet, for at least some early users, strangely compelling.

For recovering addicts, doctors at the University of Massachusetts are testing a wearable sensor designed to discern drug cravings and send text messages with just the right blend of tough love.

For those with a hankering for a custom-built companion and $125,000 to spend, a talking robotic head can be modeled on the personality of your choice. It will smile at its own jokes and recognize familiar faces.

For dieters, a 15-inch robot with a touch-screen belly, big eyes and a female voice sits on the kitchen counter and offers encouragement after calculating their calories and exercise.

"Would you come back tomorrow to talk?" the robot coach asks hopefully at the end of each session. "It's good if we can discuss your progress every day."

Robots guided by some form of artificial intelligence now explore outer space, drop bombs, perform surgery and play soccer. Computers running artificial intelligence software handle customer service calls and beat humans at chess and, maybe, "Jeopardy!"

MACHINES AS COMPANIONS

But building a machine that fills the basic human need for companionship has proved more difficult. Even at its edgiest, artificial intelligence cannot hold up its side of a wide-ranging conversation or, say, tell by an expression when someone is about to cry. Still, the new devices take advantage of the innate soft spot many people have for objects that seem to care—or need someone to care for them.

Their appearances in nursing homes, schools and the occasional living room are adding fuel to science fiction fantasies of machines that people can relate to as well as rely on. And they are adding a personal dimension to a debate over what human responsibilities machines should, and should not, be allowed to undertake.

Ms. Oldaker, a part-time administrative assistant, said she was glad Paro could keep her mother company when she could not. In the months before Mrs. Lesek died in March, the robot became a fixture in the room even during her daughter's own frequent visits.

"He likes to lie on my left arm here," Mrs. Lesek would tell her daughter. "He's learned some new words," she would report.

Ms. Oldaker readily took up the game, if that is what it was.

"Here, Mom, I'll take him," she would say, boosting Paro onto her own lap when her mother's food tray arrived.

Even when their ministrations extended beyond the robot's two-hour charge, Mrs. Lesek managed to derive a kind of maternal satisfaction from the seal's sudden stillness.

"I'm the only one who can put him to sleep," Mrs. Lesek would tell her daughter when the battery ran out.

"He was very therapeutic for her, and for me too," Ms. Oldaker said. "It was nice just to see her enjoying something."

Like pet therapy without the pet, Paro may hold benefits for patients who are allergic, and even those who are not. It need not be fed or cleaned up after, it does not bite, and it may, in some cases, offer an alternative to medication, a standard recourse for patients who are depressed or hard to control.

In Japan, about 1,000 Paros have been sold to nursing homes, hospitals and individual consumers. In Denmark, government health officials are trying to quantify its effect on blood pressure and other stress indicators. Since the robot went on sale in the United States late last year, a few elder care facilities have bought one; several dozen others, hedging their bets, have signed rental agreements with the Japanese manufacturer.

But some social critics see the use of robots with such patients as a sign of the low status of the elderly, especially those with dementia. As the technology improves, argues Sherry Turkle, a psychologist and professor at the Massachusetts Institute of Technology, it will only grow more tempting to substitute Paro and its ilk for a family member, friend—or actual pet—in an ever-widening number of situations.

"Paro is the beginning," she said. "It's allowing us to say, 'A robot makes sense in this situation.' But does it really? And then what? What about a robot that reads to your kid? A robot you tell your troubles to? Who among us will eventually be deserving enough to deserve people?"

But if there is an argument to be made that people should aspire to more for their loved ones than an emotional rapport with machines, some suggest that such relationships may not be so unfamiliar. Who among us, after all, has not feigned interest in another? Or abruptly switched off their affections, for that matter?

In any case, the question, some artificial intelligence aficionados say, is not whether to avoid the feelings that friendly machines evoke in us, but to figure out how to process them.

"We as a species have to learn how to deal with this new range of synthetic emotions that we're experiencing—synthetic in the sense that they're emanating from a manufactured object," said Timothy Hornyak, author of *Loving the Machine,* a book about robots in Japan, where the world's most rapidly aging population is showing a growing acceptance of robotic care. "Our technology," he argues, "is getting ahead of our psychology."

More proficient at emotional bonding and less toylike than their precursors—say, Aibo the metallic dog or the talking Furby of Christmas crazes past—these devices are still unlikely to replace anyone's best friend. But as the cost of making them falls, they may be vying for a silicon-based place in our affections.

Strangely Compelling

Marleen Dean, the activities manager at Vincentian Home, where Mrs. Lesek was a resident, was not easily won over. When the home bought six Paro seals with a grant from a local government this year, "I thought, 'What are they doing, paying $6,000 for a toy that I could get at a thrift store for $2?' " she said.

So she did her own test, giving residents who had responded to Paro a teddy bear with the same white fur and eyes that also opened and closed. "No reaction at all," she reported.

Vincentian now includes "Paro visits" in its daily roster of rehabilitative services, including aromatherapy and visits from real pets. Agitated residents are often calmed by Paro; perpetually unresponsive patients light up when it is placed in their hands.

"It's something about how it shimmies and opens its eyes when they talk to it," Ms. Dean said, still somewhat mystified. "It seems like it's responding to them."

Even when it is not. Part of the seal's appeal, according to Dr. Takanori Shibata, the computer scientist who invented Paro with financing from the Japanese government, stems from a kind of robotic sleight of hand. Scientists have observed that people tend to dislike robots whose behavior does not match their preconceptions. Because the technology was not sophisticated enough to conjure any animal accurately, he chose one that was unfamiliar, but still lovable enough that people could project their imaginations onto it. "People think of Paro," he said, "as 'like living.' "

It is a process he—and others—have begun calling "robot therapy."

At the Veterans Affairs Medical Center in Washington on a recent sunny afternoon, about a dozen residents and visitors from a neighboring retirement home gathered in the cafeteria for their weekly session. The guests brought their own slightly dingy-looking Paros, and in wheelchairs and walkers they took turns grooming, petting and crooning to the two robotic seals.

Paro's charms did not work on everyone.

"I'm not absolutely convinced," said Mary Anna Roche, 88, a former newspaper reporter. The seal's novelty, she suggested, would wear off quickly.

But she softened when she looked at her friend Clem Smith running her fingers through Paro's fur.

"What are they feeding you?" Ms. Smith, a Shakespeare lover who said she was 98, asked the seal. "You're getting fat."

A stickler for accuracy, Ms. Roche scolded her friend. "You're 101, remember? I was at your birthday!"

The seal stirred at her tone.

"Oh!" Ms. Roche exclaimed. "He's opening his eyes."

As the hour wore on, staff members observed that the robot facilitated human interaction, rather than replaced it.

"This is a nice gathering," said Philip Richardson, who had spoken only a few words since having a stroke a few months earlier.

Dorothy Marette, the clinical psychologist supervising the cafeteria klatch, said she initially presumed that those who responded to Paro did not realize it was a robot—or that they forgot it between visits.

Yet several patients whose mental faculties are entirely intact have made special visits to her office to see the robotic harp seal.

"I know that this isn't an animal," said Pierre Carter, 62, smiling down at the robot he calls Fluffy. "But it brings out natural feelings."

Then Dr. Marette acknowledged an observation she had made of her own behavior: "It's hard to walk down the hall with it cooing and making noises and not start talking to it. I had a car that I used to talk to that was a lot less responsive."

ACCEPTING A TRUSTY TOOL

That effect, computer science experts said, stems from what appears to be a basic human reflex to treat objects that respond to their surroundings as alive, even when we know perfectly well that they are not.

Teenagers wept over the deaths of their digital Tamagotchi pets in the late 1990s; some owners of Roomba robotic vacuum cleaners are known to dress them up and give them nicknames.

"When something responds to us, we are built for our emotions to trigger, even when we are 110 percent certain that it is not human," said Clifford Nass, a professor of computer science at Stanford University. "Which brings up the ethical question: Should you meet the needs of people with something that basically suckers them?"

An answer may lie in whether one signs on to be manipulated.

For Amna Carreiro, a program manager at the M.I.T. Media Lab who volunteered to try a prototype of Autom, the diet coach robot, the point was to lose weight. After naming her robot Maya ("Just something about the way it looked") and dutifully entering her meals and exercise on its touch screen for a few nights, "It kind of became part of the family," she said. She lost nine pounds in six weeks.

Cory Kidd, who developed Autom as a graduate student at M.I.T., said that eye contact was crucial to the robot's appeal and that he had opted for a female voice because of research showing that people see women as especially supportive and helpful. If a user enters an enthusiastic "Definitely!" to the question "Will you tell me what you've eaten today?" Autom gets right down to business. A reluctant "If you insist" elicits a more coaxing tone. It was the blend of the machine's dispassion with its personal attention that Ms. Carreiro found particularly helpful.

"It would say, 'You did not fulfill your goal today; how about 15 minutes of extra walking tomorrow?' " she recalled. "It was always ready with a Plan B."

Aetna, the insurance company, said it hoped to set up a trial to see whether people using it stayed on their diets longer than those who used other programs when the robot goes on sale next year.

Of course, Autom's users can choose to lie. That may be less feasible with an emotion detector under development with a million-dollar grant from the National Institute on Drug Abuse that is aimed at substance abusers who want to stay clean.

Dr. Edward Boyer of the University of Massachusetts Medical School plans to test the system, which he calls a "portable conscience," on Iraq veterans later this year. The volunteers will enter information, like places or people or events that set off cravings, and select a range of messages that they think will be most effective in a moment of temptation.

Then they don wristbands with sensors that detect physiological information correlated with their craving. With a spike in pulse not related to exertion, for instance, a wireless signal would alert the person's cellphone, which in turn would flash a message like "What are you doing now? Is this a good time to talk?" It might grow more insistent if there was no reply. (Hallmark has been solicited for help in generating evocative messages.)

With GPS units and the right algorithms, such a system could tactfully suggest other routes when recovering addicts approached places that hold particular temptation—like a corner where they used to buy drugs. It could show pictures of their children or play a motivational song.

"It works when you begin to see it as a trustworthy companion," Dr. Boyer said. "It's designed to be there for you."

CARI BEAUCHAMP AND JUDY BALABAN

Cary in the Sky with Diamonds

FROM *VANITY FAIR*

*Before Tim Leary, the Merry Pranksters, and the psychedelic jour-
neys of the 1960s, LSD was being used as a form of psychopharma-
cology. One place where this therapy was eagerly adopted was, as
Cari Beauchamp and Judy Balaban inform us, Hollywood.*

OUR STORY IS SET IN THE YEARS BEFORE *Mad Men*,
when Eisenhower was in the White House and America had
only 48 states. Our stage is Beverly Hills, still a small town
in 1958, where movie stars and other entertainment-industry leaders
led active but traditional, even somewhat constrained social lives.

There was a zone of privacy in that time and place we can't begin
to imagine today. Money, emotional traumas, and personal doubts
were simply not discussed, even by the closest of friends. Appear-

ances were accepted as reality, so people kept very busy making sure every aspect of their lives looked correct. That didn't mean having the most lavish house, the heftiest jewels, or the largest private plane, as it came to in later decades. It did mean dressing, behaving, and speaking appropriately; appearing to be happily married, in love, or looking for love en route to marriage; not complaining about one's career or annual income; and being enormously ambitious without evidencing any ambition whatsoever.

Social lives were just as circumspect. Dinners were small A-list gatherings at Chasen's, Romanoff's, Don the Beachcomber, or poolside barbecues at private homes. The most visible scandals arose when dancing partners who were married—but not to each other—indulged in excessive caresses or when someone (almost always a man) drank too much, though boozy belligerence and even outright drunkenness were rare to invisible.

Almost everyone smoked carton-loads of regular cigarettes, but a "joint" was a body part or a lower-class dive. If people were "doing lines," you'd have guessed they were writing screenplay dialogue or song lyrics. And if you mentioned "acid," you'd mean citrus juice or a stomach problem. Nobody in Hollywood—or almost anywhere else in the United States—had ever heard of LSD, lysergic acid diethylamide. Timothy Leary wouldn't even pop his first mushroom until 1960. So it was very out of character that against this background a group of more than 100 Hollywood-establishment types began ingesting little azure pills that resembled cake decorations as an adjunct to psychotherapy.

"When I'd say I was in therapy with a doctor using LSD, people thought I was talking about World War II landing ships"—L.S.T.'s—remembers Judy Balaban, the daughter of longtime Paramount Pictures president Barney Balaban. She didn't know much about LSD when she started taking it, in the late 50s, but, she laughingly says, "I figured if it was good enough for Cary Grant, it was good enough for me!"

If appearances were important to those behind the camera, they were crucial to stars of the big screen. And as far as the public of 1958 was concerned, Betsy Drake and Cary Grant had "perfected the ideal living pattern" after eight years of wedded bliss. According to the fan magazines, theirs had been a fairy-tale romance: Cary had seen Betsy on the London stage in 1947, and then, when they both seren- dipitously found themselves on the *Queen Mary* returning to the States, he begged a friend, the movie star Merle Oberon, to arrange an introduction. After an intense several days on shipboard, Betsy bolted into New York City, but Cary sought her out. Within months he had persuaded her to move to Los Angeles, where she signed with RKO and David O. Selznick and then burst to screen stardom op- posite Grant in *Every Girl Should Be Married*. The *Los Angeles Times* proclaimed her "the freshest, most distinctive personality since [Jean] Arthur," and Hollywood columnist Hedda Hopper declared her to be "at the threshold of a brilliant career."

Grant and Drake made headlines when they flew to Arizona to elope on Christmas Day 1949 with their pilot and Cary's best man, Howard Hughes. Betsy made a few more films before she decided to put her marriage ahead of her career. Determined to be a successful wife, she sought ways to become indispensable to a man who already had a secretary and valet. She developed into a great cook and became his trusted sounding board. She studied hypnosis and, at Cary's urging, helped both of them to stop smoking, but when he asked her to do the same for his drinking, she agreed to banish only hard liquor and not the wine and beer she enjoyed.

Betsy was beseeched for her advice on how to have a happy mar- riage, and newspapers and magazines praised the couple's simple yet complete lives, at their homes in Palm Springs and Beverly Hills or on location. She was at his side in Cannes in 1954 while he made *To Catch a Thief* with Alfred Hitchcock, and then she went to Spain to join him on the set of *The Pride and the Passion*. But it was there she realized her husband was falling in love with his co-star Sophia

Loren. When Loren came to America not long afterward to star with Grant in *Houseboat*, it was clear to Betsy that her marriage was over.

Behind the smiling pictures, Betsy was miserable. Though still in love with Grant, she tried to find the strength to leave him, but her shattered childhood had given her no psychic grounding to weather this rejection. She had been born in Paris in 1923 to wealthy parents—her grandfather had built Chicago's Drake and Blackstone hotels—and the family was living the good life in France alongside the Hemingways and other American expatriates. But following the crash of 1929 the Drakes returned to Chicago, where Betsy was ensconced at the Drake with a nanny while her parents lived at the Blackstone and worked at writing a play. They soon divorced and Betsy's mother suffered a nervous breakdown; Betsy spent the rest of her childhood being shuttled among relatives in Washington, D.C., Virginia, and Connecticut.

Without realizing it, Betsy found solace in acting; when she answered the phone pretending to be someone else, the stutter that plagued her miraculously vanished. But it wasn't until she appeared in a school play and the audience burst out in "this wonderful laughter" that she felt an approval she had never known before.

Dropping out of high school, she made the rounds of New York agents and auditions, modeling and understudying on Broadway until she was cast by Elia Kazan for a production of *Deep Are the Roots*, opposite Gordon Heath, opening in London. It was there that Cary had seen her, but taken as she was with him, she was also afraid. Betsy had had lovers before, but she resisted marriage, in large part because of what she had witnessed at home. Yet Cary was so persistent in his courting that she became convinced he was the anchor she had been seeking all her life. Twenty years her senior, he became "my lover, my husband, my everything."

With her marriage now in tatters, Betsy knew she had to talk to someone and, swearing her friend Sallie Brophy to secrecy, poured out her heart. Sallie, a stage and television actress who had suffered from depression since childhood, told Betsy that she was trying a new kind of therapy with a wonder drug that had the power to break through to the subconscious. She insisted that Betsy meet her therapist, but when they arrived at his Beverly Hills office, Betsy refused to get out of the car. So Sallie went inside and brought the doctor out. He talked to Betsy through the open car window:

"You are desperate, right?"

Betsy nodded.

"Well, then why not give this a try?"

Hardly the most persuasive argument—or the most thorough intake interview—but Betsy saw the logic and agreed to come back the next morning. She was feeling somewhat more hopeful that night when she joined Cary, Clifford Odets, and Jascha Heifetz for dinner at Chasen's. She told them, "Tomorrow I am going to take LSD." But the men looked at her blankly and then went on with their conversation. "They didn't know what I was talking about," she says. "No one had heard of it."

"I HAD A STRANGE FEELING . . ."

Twenty years earlier, in 1938, a 32-year-old Swiss chemist named Albert Hofmann had synthesized the concoction while experimenting with fungus in search of a stimulant for the central nervous system. "I had a strange feeling that it would be worthwhile to carry out more profound studies," Hofmann later said. After trying the drug himself, first by mistake and then intentionally, he added, "I became aware of the wonder of creation, the magnificence of nature."

He labeled the chemical LSD-25, because it had been the 25th variation in his experiments. His employer, Sandoz laboratories (now a subsidiary of Novartis), began providing the substance to researchers in hopes of finding profitable applications. By the mid-1950s, the C.I.A., the U.S. Army, the Canadian government, and Britain's M.I.6 had all jumped in, hoping LSD would serve as a truth serum or a new method of chemical warfare. Prisons and the military provided fertile and secret testing grounds. Other practitioners, varying widely in their legitimacy, experimented on derelicts, terminal cancer patients, residents of veterans' hospitals, and college students. Within the psychiatric profession word spread that LSD held the potential to cure alcoholism, schizophrenia, shell shock (now known as post-traumatic stress disorder), and a wide range of other problems. Between 1950 and 1965, a reported 40,000 people worldwide would be tested or "treated" with LSD.

Sandoz was so loose with its requirements for obtaining the drug that when Oscar Janiger, a Los Angeles psychiatrist, wrote the company in the mid-1950s asking for a supply to give to consenting patients, on whose experiences he would then report, he was sent his own private stock of LSD. Artists told other artists, ministers told other ministers, and the good doctor was soon spending most of his time hosting experiments. Along with Dr. Sidney Cohen, Janiger expanded his efforts into a "creativity" study through U.C.L.A., where writers, painters, and musicians such as André Previn experimented with the drug.

Aldous Huxley, the renowned author of *Brave New World* and *The Doors of Perception,* was one of the first in Los Angeles to take LSD and was soon joined by others including the writer Anaïs Nin. The screenwriter Charles Brackett discovered "infinitely more pleasure" from music on LSD than he ever had before, and the director Sidney Lumet tried it under the supervision of a former chief of psychiatry for the U.S. Navy. Lumet says his three sessions were "wonderful," especially the one where he relived his birth and, after checking with his father, learned that the experience was factually accurate, not simply symbolic. Another early experimenter was Clare Boothe Luce, the playwright and former American ambassador to Italy, who in turn encouraged her husband, *Time* publisher Henry Luce, to try LSD. He was impressed and several very positive articles about the drug's potential ran in his magazine in the late

50s and early 60s, praising Sandoz's "spotless" laboratories, "meticulous" scientists, and LSD itself as "an invaluable weapon to psychiatrists."

It was in the mid-1950s that Sallie Brophy's therapist, Mortimer Hartman, began experimenting with LSD. A radiologist, he had undergone five years of Freudian analysis and was thrilled to find a drug that seemed to let the unconscious burst to the forefront, instantaneously dissolving the ego instead of slowly peeling it away layer by layer. Claiming LSD "intensifies emotion and memory a hundred times," as Hartman told *Look* magazine in 1959, he became so enamored with the drug that he shifted away from radiology and joined forces with the psychiatrist Arthur Chandler to create the sedate yet pretentiously named Psychiatric Institute of Beverly Hills. Their next step was to secure a direct source of the drug from Sandoz for what they said would be a five-year study of LSD as a catalyst in the treatment of—as they affectionately named this new class of patients—"garden-variety neurotics."

The tall and gangly Hartman opened his institute on Beverly Hills' exclusive Lasky Drive. The rooms were furnished with sofas and decorated in what one patient remembers as "inexpensive and undistinguished browns and beiges," with wood paneling halfway up the walls. Hartman and Chandler were partners, but Chandler, whom another patient describes as looking like "an unfunny Walter Matthau," continued to work out of his house off Coldwater Canyon. In the words of a doctor who knew them both, Chandler served as a "drag" on the potentially "grandiose and messianic" Hartman, who was, after all, a doctor, but not a trained psychiatrist.

At most universities and hospitals, students and volunteers were paid for their willingness to test LSD, but Hartman and Chandler reversed the equation, and even though they saw only a few patients a day, the doctors were paid very well for their time. Aldous Huxley wrote to a friend that he found it "profoundly disturbing" to meet "two Beverly Hills psychiatrists . . . who specialize in LSD therapy at

$100 a shot—really, I have seldom met people of lower sensitivity, more vulgar mind!"

Yet the two treatment rooms at the Psychiatric Institute were soon booked five days a week after patients such as Sallie Brophy began recommending the therapy to friends such as Betsy Drake. Shown into one of the small rooms and told to lie on the couch in the corner, Betsy was given a pair of blinders to wear to block out any distractions. Assured that the tiny blue dots in the little white paper cup came straight from the Sandoz laboratories, she was soon feeling a "horrible crushing," and, in very real physical pain, she realized she was re-experiencing her own birth. The session lasted several hours and she was given a Seconal to "bring me down" slowly. Enthused by what she considered an incredible experience, Betsy went home and called her mother, with whom she hadn't spoken in more than a decade. "I told her, 'I love you,' and after all that time, she just said, 'Of course you do, darling,' and hung up."

The failure to reconnect in a meaningful way with her mother didn't dampen Betsy's optimism about the therapy. Fifty years later, sitting in her cozy London home with her bobbed hair now gray but her high cheekbones and radiant smile evidence of her long-ago stardom, she says her memories of her experiences under LSD are still crystal-clear, the revelations still vivid. The unconscious, she says, "is like a vast ocean. You don't know where you are going to go. There is no past, present, and future—all time is now. The amazing thing about the drug is the things you see. The palm trees look different. Everything looks different, and it teaches you so much."

Once a week for several months, Drake returned to Hartman's office for her sessions and her LSD, arriving at eight A.M. and staying until as late as seven at night. Like a dentist leaving a patient after administering novocaine, Hartman was in and out of the room, sometimes putting on music to enhance the atmosphere. Because it was mandated that patients not drive themselves home, friends such as Judy Balaban picked her up.

Judy was only 26, but she had been married for six years to Jay Kanter, agent to stars such as Marlon Brando, Gregory Peck, Marilyn Monroe, and Grace Kelly, who were also close friends. (Judy had served as a bridesmaid at Kelly's royal wedding, in Monaco.) Judy and Jay had two young daughters, and friends assumed her family was as perfect as it looked, but she was troubled by the sense that her life had become perfunctory, and she felt unconnected to her children. This hidden dissatisfaction with outwardly happy lives was a common theme among Betsy and Judy's circle of friends, which also included the actress Polly Bergen (recently seen on *Desperate Housewives* as Felicity Huffman's mother), who was married to agent Freddie Fields, founder of the precursor to ICM; Linda Lawson, a rising ingenue who was dating and would eventually marry the agent and future producer John Foreman (*Butch Cassidy and the Sundance Kid*); and Marion Marshall, an actress who had recently divorced the director Stanley Donen and would go on to marry the actor Robert Wagner.

In some sense, all these women were living the lives they had been raised to think they wanted. John Foreman later summarized the classic conundrum of marriages in the 1950s: "The guy rides up on a white horse, sweeps the girl off her feet, and says, 'Marry me and I'll give you everything you want.' Years pass and the wife comes to the painful conclusion that she is miserable. 'Why are you unhappy?' asks the husband. 'What do you want?' 'I don't know,' the wife responds helplessly. 'I thought you knew and were going to give it to me.'"

A few of these women had tried analysis, but none had ever been given prescriptions from their psychiatrists. Yet LSD was seen as a powerful tool to break through confusion and inhibition. As Bergen says, "I wanted to be the person, not the persona," and what attracted her to LSD therapy was "this possibility of a magic wand" that would force her to open up. Marshall, who went to Hartman's office once a week for about a year, is quick to point out that she never thought of the regimen as "taking a drug. It was therapy. It was what my doctor told me to do, so I did it."

Their descriptions of their experiences on LSD can sound today like a rehash of New Age clichés, but at the time—before the Beatles and the Jefferson Airplane were literally singing the praises of psychedelic drugs, before every college student was reading Carlos Castaneda—their perceptions were fresh and revelatory. Like Sidney Lumet and Betsy Drake, Judy relived her birth and often felt during therapy as if she had left her body and "fused" with the universe. "You experienced this otherworld consciousness and became part of what I imagined was 'the infinite mind of man.'"

Linda Lawson was unprepared when she took the little blue dots, put on her blinders, and was soon suffering "a burst of rage and sobbing." She was once again a 13-year-old girl, reliving the death of her father, "who had never raised his voice and was always so loving" but had left her to live with a mother who she felt didn't know how to love her. In grappling with her issues of abandonment, Linda grew so trusting of Hartman (she found him "sweet, if a bit skeletal") that when he urged her to move in with John Foreman she did so. And when the doctor added Ritalin—a stimulant that can affect brain chemistry—to her regimen, she didn't question him.

"MY WISE MAHATMA"

Cary Grant's initial impetus for visiting Dr. Hartman was a concern about what his wife might be saying about him. Grant methodically cultivated his debonair image and had been a leading man for more than 25 years. It was an unparalleled achievement, all the more remarkable because he had accomplished it by creating his persona out of whole cloth. He was a poor and emotionally abused boy of 14 named Archie Leach when he left his Bristol, England, home several years after his mother had simply disappeared; it would be decades before he discovered she had been institutionalized, possibly by his father, who had another family on the side. Grant came to America as an acrobat, soon began acting on the stage, and was famously "discovered" in 1932 by Mae West, who gave him his first featured film role, in *She Done Him Wrong*. He had transformed himself with a new accent and educated himself about art, clothes, and etiquette, in the process becoming the proverbial man of the world whom every woman wants and every man wants to be. He had perfected his exterior beyond his wildest dreams, but the inside was something else again. His self-deprecating remark "Everyone wants to be Cary Grant—even I want to be Cary Grant" had more than a ring of truth to it.

At the time he began treatment with Dr. Hartman he was 55 and separated from Betsy, his third wife. His first marriage, to the actress Virginia Cherrill, had lasted only a year, and his marriage to the Woolworth heiress Barbara Hutton ended after three years. (He was the only one of her eventual seven husbands not to take money from her.) Cary remained friends with Betsy, sometimes even staying with her for weekends, but Betsy was busy trying to reclaim her own life. He may not have been aware of how devastated she was by their breakup, but he did know there was a very real void in his own life.

Leery of doctors, in part because he believed Barbara Hutton's hypochondria had led to unnecessary operations and pain, Cary was not ready to be impressed with Hartman. Yet he quickly became intrigued, started calling the doctor "my wise Mahatma," and began what would be some 100 therapy sessions over several years.

There is no question that, at least for a period of time, LSD truly transformed Cary Grant. "When I first started under LSD, I found myself turning and turning on the couch," he later told a friendly reporter. "I said to the doctor, 'Why am I turning around on this sofa?' and he said 'Don't you know why?' and I said I didn't have the vaguest idea, but I wondered when it was going to stop. 'When you stop it,' he answered. Well, it was like a revelation to me, taking complete responsibility for one's own actions. I thought 'I'm unscrewing myself.' That's why people use the phrase, 'all screwed up.'"

Few of the participants mentioned their drug therapy to friends who weren't also in therapy. They did, however, talk with one another; as Judy Balaban says, "What I had with Cary and Betsy was a kind of soul-baringness that the culture didn't start to deal with until years later. We continued to have that even when our lives went off in different directions." When the actor Patrick O'Neal asked Judy about LSD during a dinner party at Oscar Levant's house, she started to explain, but Oscar interrupted with his own pithy summation: "Patrick, you don't get it. Judy was taking LSD for exactly the opposite reason you and I take stuff. She is trying to find out about things. You and I are trying to obliterate them."

Yet that was a conversation among a small group of close friends. Beyond scientific journals and mentions in *Time* magazine, there was still little information about LSD available to the public. Then, much to his friends' surprise, Cary Grant began talking about his therapy in public, lamenting, "Oh those wasted years, why didn't I do this sooner?"

This kind of sharing, as we might now call it, was very out of character for a man to whom his carefully cultivated image was so important that he had maintained more than 20 scrapbooks of the international coverage he had received. When he started taking LSD he stopped saving articles, even though there were dozens of interesting new ones he could have cut and pasted into those blank pages.

"The Curious Story Behind the New Cary Grant" headlined the September 1, 1959, issue of *Look* magazine, and inside was a glowing account of how, because of LSD therapy, "at last, I am close to happiness." He later explained that "I wanted to rid myself of all my hypocrisies. I wanted to work through the events of my childhood, my relationship with my parents and my former wives. I did not want to spend years in analysis." More articles followed, and LSD even received a variation of the Good Housekeeping Seal of Approval when that magazine declared in its September 1960 issue that it was one of the secrets of Grant's "second youth." The magazine went on to praise him for "courageously permitting himself to be one of the subjects of a psychiatric experiment with a drug that eventually may become an important tool in psychotherapy."

Many reading those articles had to be intrigued, but MGM's great aqua diva, Esther Williams, was one of the few who could pick up the phone, call Cary, and have him invite her over to discuss it. Williams had captivated audiences with her dazzling smile, her synchronized swimming, and her perfect athletic body in films such as *Million Dollar Mermaid* and *Dangerous When Wet,* but now she was in her late 30s and had just been through a wrenching divorce, only to discover that her now ex-husband had spent all her earnings and left her with a huge debt to the I.R.S. As she put it in her autobiography, "At that point, I really didn't know who I was. Was I that glamorous femme fatale? . . . Was I just another broken-down divorcée whose husband left her with all the bills and three kids?"

Now here was Cary Grant saying, "I know that, all my life, I've been going around in a fog. You're just a bunch of molecules until you know who you are." In a fog. That was exactly how Esther was feeling, and she was desperate to break through it. Cary warned her, "It takes a lot of courage to take this drug," because "it's a tremendous jolt to your mind, to your ego." After Williams assured him she had "to find some answers, fast," Grant agreed to introduce her to Dr. Hartman.

Esther, who has lived for years in Beverly Hills with her longtime husband, Ed Bell, still has a swimming pool and still remembers her experience with LSD vividly. She eagerly took her little blue pills and was thrilled to discover that "with my eyes closed, I felt my tension and resistance ease away as the hallucinogen swept through me. Then, without warning, I went right to the place where the pain lay in my psyche." She returned to the day when she was 8 years old and her beloved 16-year-old brother, Stanton, died. The family had moved from Kansas to Los Angeles, convinced Stanton was destined for stardom, and his death devastated each family member in different ways. Under LSD, Esther saw "my father's face as a ceramic plate. Almost instantly, it splintered into a million tiny pieces, like a windshield when a rock goes through it." Then she saw her mother's face on that terrible day, and "all the emotion had drained out of her, and her soft, kindly features had hardened."

During the session Esther realized—"observing it from a distance as if I were acting in or watching a movie"—that ever since the day her brother had died her life had been consumed by the necessity to replace him in every sense of the word, and "suddenly this little girl was in a race against time to be an adult."

Exhausted but calm, Esther left the doctor's office and returned to her Mandeville Canyon home, where her parents, still emotionally broken by Stanton's death, were waiting to have dinner with her. She "understood them that night in a profound way, and while I sympathized, I was also sickened by their weakness and their resignation. I saw that they both simply had given up, which, no matter what life had in store for me, was something I could never and would never do."

But the evening wasn't over for Esther. After she had said good night to her parents, she went to her bedroom, undressed, and washed. When she looked in the mirror, "I was startled by a split image: One half of my face, the right half, was me; the other half was the face of a sixteen-year-old boy. The left side of my upper body was flat and muscular. . . . I reached up with my boy's large, clumsy hand to touch my right breast and felt my penis stirring. It was a hermaphroditic phantasm." Esther has no recollection of how long she stood there, but there was no question that now "I understood perfectly: when Stanton had died, I had taken him into my life so completely that he became a part of me."

"Well, Let's Just End This"

For Esther Williams, Cary Grant, Betsy Drake, and many others, the experience of taking LSD had a profound effect on them. Over and over in interviews, former patients recounted how it changed their perception of the universe and of their place in it. Most agreed with Sidney Lumet, who says LSD provided "remarkable revelations" he continues to consider very useful to this day. Yet, in many cases, their experiences were not all positive, sometimes because of unexpected reactions to the drug, sometimes because of odd, even irresponsible actions by the therapists, who were in uncharted waters, way beyond normal medical protocols.

Marion Marshall had a frightening session where she was convinced a huge black-widow spider was going to attack her. She pulled off her mask to talk to Hartman, and when she told him what was happening, he said, "Well, let's just end this." But Marion insisted, "No, I am going to go back and face it." She put her blinders back on, and "it turned into the best session I ever had. I faced my fears, whatever they were. It was like the death experience that people describe; all of a sudden everything was white and wonderful."

She had won her revelation in spite of Hartman, who was even less helpful during what turned out to be Judy Balaban's last experience with LSD. "It started out like all my sessions," she recalls. "I went into the fusion [with the universe] state and got all the way out there, no longer connected to my body. But suddenly I hit the dysphoric side rather than the euphoric side I'd always gone to, and I was scared for the first time in eight months. I wanted to return to my body, but couldn't. I was so disconnected I couldn't even make my mouth work. Usually when you were fused, you could speak if you needed to. Not this time. After a couple of minutes of silence that felt like a year, Hartman said, 'I don't know where you are, kid . . . you're on your own!'

"You're on your own! Now I was really terrified! I'm stuck in this abstract universe, disconnected from my body, and no one knows how I can get back to myself! He gave me a shiny yellow pill—Compazine, I think—but it took several more hours for me to reconnect my body and my mind. I didn't blame Hartman for putting me there, but I did blame him for abandoning me verbally. For months afterward, usually at night, I would return to that fused state and be afraid I couldn't get back into myself. Finally, another doctor taught me how to breathe properly when an incident began, and then I was able to stop it before it took hold of me. I never had even a hint of another one again."

Polly Bergen had been going to Dr. Chandler's house once a week for several months, but when the little blue pills didn't seem to work anymore, he gave her injections of Ritalin. "Because I don't seem to have available veins elsewhere, he shot it into my hand, and when it didn't go into my veins, I watched as my hand started to swell with fluid. All the while he kept talking on and on about his own experiences. I had to tell him it wasn't working, and he took the needle out, but that's when I realized I was being treated by someone who was high, stoned, completely gone."

Having lost all confidence in Chandler, Polly stopped seeing him, but periodically she "started disappearing into this dreamlike state, not actually leaving my body, but reliving these experiences: being born, being a child in a crib." The flashbacks scared her, and they didn't stop until she and her husband sat down with another psychiatrist, who explained the drug and its effects, something Chandler had never done.

Linda Lawson kept trying to see the positive side of her treatments until, during one of her sessions, she heard the tinkling of glass. She lifted her blinders to see where the noise was coming from and saw Chandler "playing with these pieces of glass, making a mosaic. He was stoned and just somewhere else entirely." That did it for Linda, but occasionally she would visit him "just to sit up and talk," concluding that "he was probably a very good therapist before he started getting so stoned himself."

"Too Much of a Good Thing"

Betsy Drake credits LSD therapy with "giving me the courage to leave my husband" and, for the first time, to truly speak her mind. "After an LSD session, one morning in bed while we were both having breakfast, Cary asked me a question and I said, 'Go fuck yourself.' He jumped out of bed, buttoning the top of his pajamas, his bare bottom showing, and slammed the bathroom door. That was the true beginning of the end."

She and Cary were divorced in 1962 after 13 years of marriage—his longest—but they remained friendly for the rest of his life. The therapy had intensified her interest in the mental-health field; she began volunteering, then studying, at U.C.LA.'s Neuropsychiatric Institute and other Los Angeles hospitals. In the early 70s she published a novel and enrolled at Harvard, earning a master's of education in psychology, specializing in psychodrama therapy, where patients act out problems instead of discussing them.

Cary continued to sing the praises of LSD, and his belief in it was evidenced by the fact he left Dr. Hartman $10,000 in his will. But when the actress Dyan Cannon divorced Grant in 1968, after less than three years of marriage, LSD was used against him. In seeking custody of their daughter, Jennifer, Cannon's lawyers claimed that he was "an unfit father" because of his use of the drug and his resulting "instability." However, when the respected psychiatrist Judd Marmor testified that Grant had told him LSD had deepened the actor's "sense of compassion for people, deepened his understanding of himself, and helped cure his shyness and anxiety in dealing with other people," Grant was given two months a year with his daughter and the right to overnight visits.

Grant's defensive posture regarding LSD during his last divorce reflected the dramatic shift in public opinion. Beginning in 1962, the Food and Drug Administration began demanding to see the records of doctors such as Hartman and Chandler and appeared at their offices to confiscate their LSD supply. The doors of the Psychiatric Institute of Beverly Hills closed suddenly that same year. Linda Lawson remembers being deep into her drug-induced state when Hartman informed her, without giving any reason, that he was leaving California and this would be her last session with him. The proliferation of LSD as a street drug and reports of suicides and other tragic consequences of LSD abuse led to national legislation criminalizing its possession in 1968. There wasn't much resistance from its earliest adherents. Clare Boothe Luce was said to have cautioned, "We wouldn't want everyone doing too much of a good thing."

Nevertheless, one of the common threads among the interviews we conducted with past patients was that, no matter how they felt about their personal experience with LSD, they resented that Timothy Leary's much-publicized campaign to "turn on, tune in, drop out" had sparked a backlash against a drug they still believe to be a potentially beneficial telescope into the subconscious. Their time might have finally come, for today, after 50 years of its being demonized, LSD is beginning to make a comeback in the laboratory. No breakthroughs are expected soon, but researchers from around the world gathered in California this past April to compare notes, and scientists at Harvard and the University of California at San Francisco have received permission from the F.D.A. to experiment with LSD once again.

JOHN BRENKUS

The Longest Home Run Ever

FROM *THE WEEK*

> *They say that baseball is a game of inches. As John Brenkus, host of ESPN's* Sport Science, *explains, science reveals that even finer measurements are involved.*

T HAT ANYONE CAN EVEN HIT A BIG-LEAGUE PITCH is a wonder in itself. That some can hit home runs is practically a miracle. On paper, at least, the feat seems impossible.

A pitcher starts his windup for each pitch at a distance of 60 feet six inches from home plate. But by the time he releases the ball, he's about five feet closer to the plate. If he throws a 99-mph fastball, the ball is going to reach the batter in 395 milliseconds. By comparison, it takes 400 milliseconds to blink your eye completely.

A lot has to happen in those 395 milliseconds. It takes the first 100 just for the batter to see the ball in free flight and get an image of it to his brain. If a decision is made to swing, the batter generally has a grand total of 150 milliseconds to get the bat around and through the strike zone.

And those are only for the gross movements involved. There's still some fine-tuning to do. If the batter is only seven milliseconds early or late in connecting with the ball, he's going to send it foul. And even if his timing is perfect, he still has to put the "sweet spot" of the bat within an eighth of an inch of the correct spot on the ball. To give you an idea of the margin of error, the width of an average pencil is *twice* as big as the margin of error on a major league bat.

To top it off, the batter has to swing pretty hard. If he's going to hit a home run, he has to swing *very* hard, and as every golfer, tennis player, and place-kicker knows, the harder you try to hit, the tougher it is to hit with accuracy.

If you told all of this to an alien freshly landed from Mars, he'd refuse to believe that anyone has ever hit a home run, except maybe by pure luck once every 25 years or so. And he's never going to believe that Mickey Mantle hit a 565-foot bomb—at Griffith Stadium in Washington, D.C., on April 17, 1953. That shot remains baseball's official record holder.

It's natural to think about human limits. So what is the longest home run it's possible to hit?

FIVE HUNDRED FOOT HOME RUNS are still extremely rare. But to predict exactly how much farther beyond 500 feet a homer can travel, we first need to know what kind of pitch produces the longest ball. Simple physics tells us that we want the fastest pitch possible, because the ball's forward energy will be returned when it bounces off the face of the bat. To prove this to yourself, imagine throwing a ball against a wall: The harder you throw it, the faster it's going to bounce back at you.

That would seem to complicate things a little, because in order to figure out the longest possible homer, we also have to figure out the fastest possible pitch. The good news here is that we don't need to worry too much about that because, unlike many athletic skills, pitching seems to have a fairly definite outer limit.

According to Dr. Bassil Aish, chief medical advisor for the ESPN show *Sport Science*, the limiting factor is not muscle power or technique. It's how hard a pitcher could throw without dislocating his shoulder or tearing a rotator cuff or pulling a tendon off a bone. Aish has calculated that an appropriately proportioned pitcher could get strong enough to throw well above 105 mph—the fastest pitch ever recorded in the major leagues—but that anything over 111 is almost sure to result in serious damage.

If you think serious injury as a result of throwing a hard pitch is just theoretical, just ask Tony Saunders of the Tampa Bay Devil Rays, who broke his humerus, the bone that runs from the shoulder to the elbow, on a 3–2 pitch to Juan Gonzalez of the Texas Rangers in 1999. Of course, Saunders and the other major leaguers who've suffered similar injuries weren't throwing anywhere near 111 mph, which is what we're going to use for the pitch that would produce the longest possible home run.

IT SEEMS SIMPLE NOW: JUST launch a 111-mph missile and let the batter hit it with everything he's got at an optimum launch angle. But it isn't that simple, because there's air friction involved. That's going to slow the ball down, and while there's not a lot we can do about that, there is a little. As long as we're going to encounter friction, we might as well use it to our advantage. We do that by imparting some backspin to the ball when it's hit, changing its trajectory in a way that will increase distance.

It's not easy to hit a 111-mph pitch, but we're not going to concern ourselves with that difficulty: After all, our guy has to hit it only once. All we care about is the type and speed of the pitch when it does get hit, and the pitch we want is a fastball.

There are other factors that come into play. For one thing, we want the driest ballpark possible, because the "springiness" of the ball decreases as the air gets more humid, resulting in less distance. Temperature counts, too. Cold air is denser and puts more drag on the ball. When the air temperature is 75 degrees or higher, about 4 percent of batted balls result in homers. When it's colder, that number can drop to as low as 3.2 percent. To decrease air density even further, we want a stadium at high elevation. Wind speed enters the equation as well. But for our longest-ball prediction, we're going to assume a windless day because we're trying to calculate the ultimate human performance under repeatable conditions.

All we need now is the fastest possible bat speed.

THE EASIEST WAY TO TALK about what we're looking for is to start with someone we know already who has one of the highest measured bat speeds in the game. That would be Derek Bell, an outfielder who had an 11-year major league career highlighted by his 1995 season, when he hit .334 for the Houston Astros. Bell was 6 foot 2 at 215 pounds in his playing days, and his highest recorded bat speed was an incredible 95 mph. Using this as a starting point, we'll break the swing into its major components to analyze how that speed could be exceeded.

According to Dr. Aish, a batter's hands, wrists, elbows, and shoulders form a "kinetic chain" enabling the linked sequence of motions that results in rotational acceleration of the upper extremities. At the "extreme of the extremities" there's a baseball bat, a highly leveraged extension of the body. The purpose of all that rotating is to get the bat's sweet spot moving through the air as fast as possible.

One way our ideal hitter could increase bat speed is simply by having longer arms. Swing a rock tied to a 1-foot string around your head at one rotation per second and the rock will move at about 4 mph. But make it a 2-foot string, and the same rotational speed will get the rock moving twice as fast. Similarly, for any given rpm, the farther you can get the bat from the center of rotation, which is the center of the batter's body, the faster the sweet spot of the bat will be moving when it meets the ball. A player 6 feet 8 inches tall would have a swing arc about 15 percent longer than Derek Bell's for the same given bat size. This translates to a bat speed of 109 mph based on arc size alone.

Of course, that extra speed is not free: You have to add more power in order to maintain the same rotational speed with longer arms and a longer bat. So the second thing we need to do is give our batter the power to increase the rotational speed of his upper extremities. An increase in muscle mass without counterproductive loss of flexibility would be about 20 percent over Derek Bell's playing weight of 215 pounds and estimated body fat of 9 percent.

The lower body contributes torque to the upper body, and by Aish's estimates, the right physique—with no performance-enhancing drugs allowed—would add 4 percent more power. He also believes that a quick, well-timed forward stride by a 6-foot-8 man can add up to 7 mph to bat speed, bringing us up to 120.6 mph.

While body rotation is doing all its work to get the bat accelerating to as high a speed as possible, there's still one more opportunity to impart a last burst of power—the "snap of the wrists" that comes at the last instant before contact. That last snap can account for as much as 20 percent of the ultimate bat speed. In technical terms, the wrist slides from maximum radial deviation to maximum ulnar deviation, and because many athletes have unusually high radial/ulnar deviations, we're going to assume that our ideal hitter has a maximum deviation of 115 degrees. (The structure of the bones precludes anything higher.) That gets 25 percent more power out of the wrist snap than Bell achieved. Twenty-five percent more of a 20 percent contribution to bat speed gets us an overall increase of 5 percent.

Final bat speed? Nearly 127 mph.

* * *

WE NOW HAVE EVERYTHING WE need to calculate the perfection point for the longest ball it's possible for a human batter to hit. It's not an easy calculation. It involves a series of differential equations developed by Robert Adair for his book *The Physics of Baseball.*

But imagine that 400 years from now, a slugger named Smith will step to the plate one warm day during a game at Coors Field in mile-high downtown Denver. Smith, who will stand 6 foot 8 and weigh 247 pounds, will be facing a rookie flamethrower fresh out of the bullpen. On the fifth pitch of the at-bat, the rookie will tilt back and unleash a 111-mph fastball over the heart of the plate.

No athlete uses more of his muscles at one time than a batter, and Smith will be able to feel his coming into perfectly orchestrated play as he starts the bat around, pours power into it, and keeps pouring it on through the entire arc of his swing.

The ball will have slowed to 100 mph by the time the bat, moving at 127, finally connects. The sound of impact will be like nothing anyone in the ballpark has ever heard.

Even in the upper deck it will sound crisp and sharp, like a rifle shot or a dry log breaking cleanly in two.

The ball will leave the face of the bat at 194 mph and soar upward at a 35-degree angle. Backspin will cause it to rise sharply at first, and it will still be heading upward when it rises above roof level.

When the ball finally lands, 9.3 seconds after Smith hits it, it will strike a patch of dirt outside the stadium and leave a sharp indentation. After the game, measurements will be taken that show the ball traveled exactly 748 feet from home plate.

Will it happen? In reality, probably not. Because the danger to pitchers would be far too great, it's likely that baseball's rules or equipment would be changed before a ball ever flew that far. But it could.

ALAN SCHWARZ

Professor Tracks Injuries with Aim of Prevention

FROM *THE NEW YORK TIMES*

Increased scrutiny into the effects of head injuries has created a new understanding of how violent—and even deadly—sports can be. Alan Schwarz profiles the man whose decades of research has helped prevent catastrophic injuries.

CHAPEL HILL, N.C.—THE MAN WITH PERHAPS THE most gruesome job in sports was unenviably busy. While other football fans spent the last weekend of October watching games, the 74-year-old retiree prepared still more formal inquiries into events that occupy him more than anyone would prefer—two high school football tragedies.

He gathered information from Web searches and e-mailed questionnaires to the National Federation of State High School Associations. The linebacker outside Kansas City, Kan., who collapsed from an apparent brain injury and died the next morning. The junior-varsity defensive back from Fresno, Calif., who was sent to a hospital and into a coma by a hit that caused massive brain swelling.

Fred Mueller has almost singlehandedly run the National Center for Catastrophic Injury Research at the University of North Carolina for 30 years, logging and analyzing more than 1,000 fatal, paralytic or otherwise ghastly injuries in sports from peewees to the pros. His work has repeatedly improved safety for young athletes by identifying patterns that lead to changes in rules, field dimensions and more.

Professional football spent most of October wrestling with how to distill illegal head-to-head collisions out of the sport's Newtonian chaos. But when Mueller calmly affirms with a nod from behind his desk that this fall's football catastrophic log is in fact no longer than usual, the knowledge that it used to be worse is somewhat hollow consolation.

"When you see a kid's picture in the paper and he's either dead or paralyzed for the rest of his life, it can get to you," said Mueller, a retired professor at Chapel Hill who still spends most of the workweek on this research. "Or when you have to call the coach or the athletic trainer to learn more about what happened. What I try to look at is, gathering the data is maybe going to prevent the same thing from happening in the future."

Mueller has no ghoulish tendencies or antipathy toward sports—in fact, he played both sides of the Tar Heels' line from 1958 to 1960, and he still works out daily in the student gym and bicycles 15 miles during lunch hours. Rather, he has witnessed the power of this data, how gathering these pixels of isolated grief can form pictures of progress.

Begun in 1931 by the American Football Coaches Association, the football death log started to be overseen by the University of North Carolina in the 1960s, when Mueller was a football and lacrosse coach willing to chip in some hours. He never left, and soon after becoming the center's director in 1980, he expanded it to include catastrophic injuries in all sports, among boys and girls.

Mueller almost immediately noticed a previously hidden cluster in, of all things, pole-vaulting. Several high school and college athletes each year were killed or paralyzed simply by missing the pit with the pole, falling on their heads off the landing pad, or sliding down the pole and hitting their heads on hard surfaces. Pits were soon expanded and surrounded with softer padding.

Mueller detected a strange number of paralytic accidents in organized swimming, all from relay-type dives into water that was too shallow—resulting in today's minimum depths. He was stunned at the number of injuries among cheerleaders, specifically those who are thrown up to 25 feet high and not caught, which led to other reforms.

"Fred is really dedicated," said Dr. Barry Boden, an orthopedic surgeon in Rockville, Md., who has used the injury logs to conduct research of his own. "We've looked at wrestling, cheerleading, baseball, football. He's very generous and a real team player. You can't find this data anywhere else."

Gathering the information was once quite scattershot. An expensive newspaper clipping service probably found most athlete deaths but surely missed numerous nonfatal cases involving brain hemorrhages or paralysis. ("Those who recovered usually didn't make the papers," Mueller said.) Web news alerts and other search techniques have recently provided a fuller picture of the rare but real risks of organized sports.

And now, people find Mueller within hours of accidents, usually those involving football. A Kansas City–area radio station called him Oct. 29 after Nathan Stiles, a Spring Hill High School linebacker, died from what appeared to be second-impact syndrome— in which a child's playing too soon after one concussion can allow a subsequent head impact to cause enormous intracranial swelling.

Mueller shared statistics on where the Stiles case fit into the larger landscape. Stiles's was the first direct death among high school players this year, compared with a typical total of about three. It was also the third catastrophic football injury in 13 days, after the paralysis of Eric LeGrand of Rutgers and Christopher Norton of Luther College in Iowa.

Within hours, Mueller learned of a fourth—the boy from Fresno. Typically, there are 36 of those each year; this year's running total is 24.

Mueller prepared the standard requests for information: the athlete's birthday, how the injury occurred, the playing surface, helmet details if applicable, and so on. Most are returned within several months. Occasionally Mueller will have to call the school's athletic trainer or coach and proceed gingerly. He does not call the parents, though they sometimes call him.

"I think sometimes people at schools are concerned they'll be blamed for the injury or death, or that it won't be good for the sport," Mueller said. "But you explain that names are all confidential, and it's only for analysis to make sports safer. History proves that."

Mueller will present that history in his book *Football Fatalities and Catastrophic Injuries, 1931–2008* (Carolina Academic Press), written with Dr. Robert Cantu, who is the Chapel Hill center's medical director and a primary voice on athletic brain injuries. More than 250 pages detail football's decade-by-decade tragedies and rule changes—like the 1976 outlawing of spearing and more recent adjustments to kickoff wedges. A final chapter discusses injury prevention strategies and other ways to make football safer.

As that book goes to press, Mueller continues to take his phone calls and scours the Web alongside file cabinets that read "Football Fatality Reports" and "Cat. Cases," short for catastrophic. He seeks the stories nobody wants to hear, the most gruesome job in sports.

"You could look at it that way," Mueller said. "But you can also look at it as the best. You're preventing deaths and disability injuries. That can be pretty satisfying."

Deborah Blum

The Trouble with Scientists

FROM THE *SPEAKEASY SCIENCE* BLOG

If scientists are suspicious of journalists and bloggers, Deborah Blum argues, then they should consider how else they can reach a wider audience to promote greater scientific literacy.

WHEN I FIRST STARTED IN JOURNALISM I WORKED as a general assignment reporter. After a few years, I decided to become a science journalist. I thought it made sense, to focus on a subject that fascinated me rather than rattling indefinitely from one news beat to another. I still remember the look of frozen horror on my father's face when I announced the decision.

As you may deduce, my father, Murray S. Blum, is a scientist. He received his PhD in 1955 from the University of Illinois, where in addition to studying entomology he was taught the essential lesson that "real" scientists shared their work only with each other and did not attempt to become "popularizers" because that would lead to "dumbing down" the research.

· He emerged from paralysis to say: "I hope you're not planning to interview my friends." A science historian at the California Institute of Technology once told me that this disdain is rooted in the way we teach science. In particular, he said, K–12 science classes in the United States are essentially designed as a filtration system, separating whose fit for what he called "the priesthood of science" from the unfit rest of us.

"Why would I want to interview boring old entomologists?" I naturally replied. This conversation was in my parents' living room (father in armchair, daughter pacing) but variations on this theme occur any time, any place. Scientists won't talk to journalists; they don't want to waste their time "dumbing it down"; they don't see it as "making us smarter." So many of the good stories in science don't get covered at all. Or the stories get covered only for an already science-literate audience—explored in publications like *Discover* or *Science News*—rather than for that far larger group, the science disenfranchised.

A recent essay by the editor of *Analytical Chemistry,* Royce Murray of the University of North Carolina, inspired in me that old urge to pace in my parents' living room. Published in October 2010 and titled "Science Blogs and Caveat Emptor," the editorial warned scientists to be wary of trusting bloggers to be fair, accurate or educated about science. In fact, he mourned the passing of good-old traditional journalism, reminding me that while the medium may change, the attitude remains the same.

I'm far from the only blogger to find this wrong-headed, of course. David Kroll, a colleague of mine at the Public Library of Science blog network, pointed out that, in general, science bloggers tend to be remarkably respectable, ranging from award-winning journalists to scientists themselves. Kroll belongs to the latter group; he just happens to be chair of the pharmaceutical sciences department at North Carolina State University. His elegantly precise counter to Murray's broad-brush declaration that "the current phenomenon of 'bloggers' should be of serious concern to scientists" made an elegantly precise point: the editor should have done his homework first.

My first reaction to Murray's piece was to wonder if he belonged to my father's generation of scientists-who-hate-to-share. Sure enough, I discovered that Murray received his PhD in 1960. In other words, we'll really move forward in improving public understanding of science, when the research community moves closer to Kroll's kind of 21st century mindset.

For one thing, one of Kroll's remedies is to suggest that more scientists become bloggers—yes, public communicators of science—themselves. I've always thought that my own profession of science journalism grew to fill the void created by scientists who couldn't be bothered to "dumb down" their work. Since the mid-1950s, the National Association of Science Writers (and, yes, I'm a past-president so I like to mention it) has grown from several hundred members to nearly 3,000. At the same time, science journalism programs have sprung up at universities from UC-Berkeley to New York University.

Science writers, journalists, broadcasters and bloggers became the voice of science during a time during which too many scientists simply refused to engage. Scientists have ceded that position of power amazingly readily; ask yourself how many research associations offer awards to journalists for communicating about science but none to their own members for doing the same? Ask yourself how the culture of science responds even today to researchers who become popular authors or bloggers, public figures. Whether young scientists are rewarded for spending time on public communication? And ask yourself how hypocritical this is, to complain that the general public doesn't understand science while refusing to participate in changing that problem?

As it turns out, the culture of the "real" scientist who exists somehow separate from the rest of us has not been a boon for public understanding or appreciation of science. So let me make a case that it's not too late for Prof. Murray and those who think like him to approach science communication differently. And it doesn't hurt to remember that we in the science-literate section of the bleachers aren't the only ones who matter here. Murray writes that he's worried about the anti-science voices on the Internet. Let me suggest that the best way to remedy that concern is probably not through an inner-circle rant in *Analytical Chemistry,* but by sowing really good information across a very much broader field of interest.

To end on a happy note, though, my father decided not to disown me after all, that having a science journalist daughter wasn't quite as embarrassing as he'd anticipated. He started calling his friends to make sure they would talk to me. He went on the *Today* show and persuaded former host Bryant Gumbel to eat beetles (flavored with a little A-1 Sauce). Of course, he once gave an interview to the *National Enquirer,* under the impression that it was the *National Observer.* Somehow, *Enquirer* readers were left with the impression that he wanted to attack enemy armies with crazed honeybees. This, um, slightly misrepresented his position. But as I keep telling him, he should congratulate himself on reaching a new—and very large—audience.

ED YONG

Gut Bacteria in Japanese People Borrowed Sushi-Digesting Genes from Ocean Bacteria

FROM *DISCOVER*'S *NOT EXACTLY ROCKET SCIENCE* BLOG

Japanese people may be genetically adapted to eating sushi. But, as Ed Yong explains, what's more remarkable is how those genes were acquired.

JAPANESE PEOPLE HAVE SPECIAL TOOLS THAT LET them get more out of eating sushi than Americans can. They are probably raised with these utensils from an early age and each person wields millions of them. By now, you've probably worked out that I'm not talking about chopsticks.

The tools in question are genes that can break down some of the complex carbohydrate molecules in seaweed, one of the main ingredients in sushi. The genes are wielded by the hordes of bacteria lurking in the guts of every Japanese person, but not by those in American intestines. And most amazingly of all, this genetic cutlery set is a loan. Some gut bacteria have borrowed their seaweed-digesting genes from other microbes living in the coastal oceans. This is the story of how these genes emigrated from the sea into the bowels of Japanese people.

Within each of our bowels, there are around a hundred *trillion* microbes, whose cells outnumber our own by ten to one. This "gut microbiome" acts like an extra organ, helping us to digest molecules in our food that we couldn't break down ourselves. These include the large carbohydrate molecules found in the plants we eat. But marine algae—seaweeds—contain special sulphur-rich carbohydrates that aren't found on land. Breaking these down is a tough challenge for our partners-in-digestion. The genes and enzymes that they normally use aren't up to the task.

Fortunately, bacteria aren't just limited to the genes that they inherit from their ancestors. Individuals can swap genes as easily as we humans trade money or gifts. This "horizontal gene transfer" means that bacteria have an entire kingdom of genes, ripe for the borrowing. All they need to do is sidle up to the right donor. And in the world's oceans, one such donor exists—a seagoing bacterium called *Zobellia galactanivorans*.

Zobellia is a seaweed-eater. It lives on, and digests, several species including those used to make nori. Nori is an extremely common ingredient in Japanese cuisine, used to garnish dishes and wrap sushi. And when hungry diners wolfed down morsels of these algae, some of them also swallowed marine bacteria. Suddenly, this exotic species was thrust among our own gut residents. As the unlikely partners mingled, they traded genes, including those that allow them to break down the carbohydrates of their marine meals. The gut bacteria suddenly gained the ability to exploit an extra source of energy and those that retained their genetic loans prospered.

This incredible genetic voyage from sea to land was charted by Jan-Hendrik Hehemann from the University of Victoria. Hehemann was originally on the hunt for genes that could help bacteria to digest the unique carbohydrates of seaweed, such as porphyran. He had no idea where this quest would eventually lead. Mirjam Czjzek, one of the study leaders, said, "The link to the Japanese human gut bacteria was just a very lucky, opportunistic hit that we clearly had no idea about before starting our project. Like so often in science, chance is a good collaborative fellow!"

FROM OCEANS TO BOWELS

Hehemann began with *Zobellia*, whose genome had been recently sequenced. This bacterium turned out to be the proud owner of no fewer than five porphyran-breaking enzymes. These enzymes were entirely new to science, they are all closely related and they clearly originated in marine bacteria. Their unique ability earned them the name of "porphyranases" and the genes that encode them were named PorA, PorB, PorC and so on.

They are clearly not alone. Using his quintet as a guide, Hehemann found six more genes with similar abilities. Five of them hailed from the genomes of other marine bacteria—that was hardly surprising. But the sixth source was a far bigger shock: the human gut bacterium *Bacteroides plebeius*. What was an oceanic gene doing in such an unlikely species? Previous studies provided a massive clue. Until then, six strains of *B. plebeius* had been discovered, and all of them came from the bowels of Japanese people.

Nori is, by far, the most likely source of bacteria with porphyran-digesting genes. It's the only food that humans eat that contains any porphyrans and until recently, Japanese chefs didn't cook nori before eating it. Any bacteria that lingered on the green fronds weren't killed before they could mingle with gut bacteria like *B.plebius*. Ruth Ley, who works on microbiomes, says, "People have been saying that gut microbes can pick up genes from environmental microbes but it's never been demonstrated as beautifully as in this paper."

In fact, *B. plebeius* seems to have a habit of scrounging genes from marine bacteria. Its genome is rife with genes that are more closely related to their counterparts in marine species like *Zobellia* than to those in other gut microbes. All of these borrowed genes do the same thing—they break down the complex carbohydrates of marine algae.

To see whether this was a common event, Hehemann screened the gut bacteria of 13 Japanese volunteers for signs of porphyranases. These "gut metagenomes" yielded at least seven potential enzymes that fitted the bill, along with six others from another group with a similar role. On the other hand, Hehemann couldn't find a single such gene among 18 North Americans. "We were trying at lunch to think about where you might see patterns this clean," says Ley. "You'd have to find another group of people with a very specialized diet. Because this involved seaweed and marine bacteria, it might be one of the cleanest demonstrations you'd get."

For now, it's not clear how long these marine genes have been living inside the bowels of the Japanese. People might only gain the genes after eating lots and lots of sushi but Hehemann has some evidence that they could be passed down from parent to child. One of the people he studied was an unweaned baby girl, who had clearly never eaten a mouthful of sushi in her life. And yet, her gut bacteria had a porphyranase gene, just as her mother's did. We already know that mums can pass on their microbiomes to their children, so if mummy's gut bacteria can break down seaweed carbs, then baby's bugs should also be able to.

Are We What We Eat?

This study is just the beginning. Throughout our history, our diet has changed substantially and every mouthful of new food could have acted as a genetic tasting platter for our gut bacteria to sample. Personally, I've been eating sushi for around two years now and I was intrigued to know if my own intestinal buddies have gained incredible new powers since then. Sadly, Czjzek dispelled my illusions. "Today, sushi is prepared with roasted nori and the chance of making contact with marine bacteria is low," she said. The project's other leader, Gurvan Michel, concurs. He notes that of all the gut bacteria from the Japanese volunteers, only *B. plebeius* as acquired the porphyranase enzymes. "This horizontal gene transfer remains a rare event," he says.

Michel also says that for these genes to become permanent fixtures of the *B. plebeius* repertoire, the bacterium would have needed a strong evolutionary pressure to keep them. "Daily access to ingested seaweeds as a carbon source" would have provided such a pressure. My weekly nibbles on highly sterile pieces of sushi probably wouldn't.

That's one question down; there are many to go. How did the advent of agriculture or cooking affect this genetic bonanza? How is the Western style of hyper-hygienic, processed and mass-produced food doing so now? As different styles of cuisines spread all over the globe, will our bacterial passengers also become more genetically uniform?

The only way to get more answers is to accelerate our efforts to sequence different gut microbiomes. Let's take a look at those of other human populations, including hunter-gatherers. Let's peer into fossilized or mummified stool samples left behind by our ancestors. Let's look inside the intestines of our closest relatives, the great apes. These investigations will tell us more about the intestinal genetic trade that has surely played a big role in our evolution.

Rob Knight, a microbiome researcher from the University of Colorado, agrees. "This result reinforces the need to conduct a broad and culturally diverse survey of who harbors what microbes. The key to understanding obesity or IBD might well be in genes or microbes acquired under circumstances very different to those we experience in Western society." Gastronomics, anyone?

Reference: Hehemann, J., Correc, G., Barbeyron, T., Helbert, W., Czjzek, M., & Michel, G. (2010). Transfer of carbohydrate-active enzymes from marine bacteria to Japanese gut microbiota. *Nature, 464* (7290), 908–912 DOI: 10.1038/nature08937

The Data Trail

FROM *OnEarth*

Tim Folger meets a retired probation officer whose weekly hikes in the Sonoran Desert have produced a remarkable set of data that researchers are using to tease out the effects of climate change on this complex ecosystem.

THERE'S SOMETHING DIFFERENT ABOUT THE DESERT this morning. Something's missing. I don't notice it at first, but my companion, who has hiked in the Sonoran Desert every week for nearly 30 years, stops on the trail ahead of me and cocks his head.

"Listen . . . not a single bird," Dave Bertelsen says. "We should be hearing cactus wrens, canyon wrens, curve-billed thrashers, *Phainopepla*"—a crested desert songbird—"Gambel's quail, Gila woodpeckers. Even in the dead of winter there are birds. This is to-tally unique. We should be able to just walk along talking and hear birds. To stop and listen hard—I've never had to do that before."

We're climbing a winding path on a rock-strewn slope in Saguaro National Park, a few miles west of Tucson's city limits. The sun, just four days shy of the winter solstice, will be rising soon. As the world pirouettes out of darkness, a diffuse pink light hides the stars and temporarily softens a hushed landscape in which almost everything seems to be barbed, sharp, or hard. In the still, cool air, a hundred million giant saguaro cacti from here to northern Mexico brace for the dawn, getting a few last gulps of carbon dioxide before sealing their pores and holding their breath all day long to minimize water loss.

Bertelsen doesn't know what to make of the absence of birds on this mid-December morning. For now it's another datum, brand new, puzzling, and disturbing. Besides, we're not on his favorite trail, north of the city in the Santa Catalina Mountains, the one he has walked 1,270 times—and counting—since 1981. During that span Bertelsen has amassed an enormous amount of information on the elevation, distribution, and bloom dates of some 600 plant species and subspecies; in 1997 he began keeping equally detailed records of the reptiles and mammals he has encountered during his weekly 10-mile hikes. Last year he added birds. "I now have 195,000 observations," he tells me as we saunter among saguaros, some of them as tall as four-story buildings. "It's a pretty substantial data set."

The decades spent walking this landscape have made the 68-year-old retired probation officer a leading expert on the Sonoran Desert's unique flora and fauna. Bertelsen's mile-by-mile notes of his treks are so precise and voluminous that a team of scientists at the University of Arizona in Tucson is using them to study the effects of global warming here. His records clearly show that about 25 percent of the plant species he has tracked have shifted their ranges to higher, cooler elevations, a response to desert summers that are now close to

2 degrees Fahrenheit warmer than they were 20 years ago. The change is significant, but Bertelsen worries more about stasis.

"To me what's interesting is not the 25 percent of plants that have adapted by moving up. It's the 75 percent that are not moving up," Bertelsen says. "Twenty-five percent is a lot, but 75 percent *aren't* adapting. That has big implications. It means most of the desert is not adapting to climate change. Since I started my hikes, the flora have declined 19 percent—that's species in bloom per mile that I actually see when I'm hiking. For fauna it's a 43.5 percent decline per mile. We're going to lose a tremendous amount."

Compared with other besieged but more luxuriant ecosystems, deserts might seem to be relatively hardened to damage, harsh places inhabited by species already used to living on the edge. What, after all, could it matter if a desert, of all places, becomes a little warmer?

By one definition the Sonoran Desert isn't a desert at all. With 11 or 12 inches of rainfall in a good year, parts of it can exceed the 10-inch limit sometimes used to designate a desert. More generally, though, a desert is defined as a region where water scarcity imposes drastic constraints on life, and the Sonoran Desert easily meets that criterion. It covers approximately 100,000 square miles, from southern Arizona and southeastern California to Mexico's northwestern coast, including most of the Baja peninsula. Of North America's four deserts, the Sonoran contains by far the greatest diversity of plant and animal species. Unlike the Mojave and Great Basin deserts to the north and the Chihuahuan Desert to the east—which all have cold winters and one rainy season—the Sonoran has mild winters and two rainy seasons, one resulting from winter storms in the Pacific and another from summer monsoons that blow in from the Gulf of California.

Without that second pulse of moisture, the Sonoran Desert would blend almost seamlessly with the continent's other deserts. Low shrubs would dominate the terrain; some annuals would bloom in exceptionally wet years; trees would be scarce. Instead, the extra rain nurtures life found nowhere else in the world. Saguaros, the iconic cacti with great upraised arms, grow only here, along with more than 2,000 other plant species. More than 350 bird species, 60 kinds of mammals, 100 different reptiles, 30 types of freshwater fish, and hundreds of thousands of invertebrates live in the Sonoran Desert.

A winter storm watered the desert a few days before my first hike with Bertelsen, and it shows, if you know how to look. Saguaros, like everything here, have evolved to take maximum advantage of intermittent rains. The trunk and arms of a saguaro have vertical pleats, so the entire cactus can inflate like a bellows and store the water absorbed by its roots, which lie just three inches or so below the surface. The roots spread to a distance about equal to the height of the cactus and can guzzle 200 gallons from one rainfall, liquid life that will sustain a saguaro for a year.

"This one is full of water," Bertelsen says, pointing with one of his walking poles to a 30-foot-tall saguaro. The waxy surface of the tumescent cactus has become smooth and even.

THE SUN ASCENDS WITH US as we continue up the slope of what was, 65 million years ago, the caldera of a volcano. Although the morning remains cool, not more than 60 degrees yet, there is no shelter from the sun. Bertelsen moves at a careful, steady pace, though he's a bit slower now, he says, than before his triple-bypass surgery in 2004, the same year in which he broke a leg and had to be helicoptered off his favorite mountain trail one night. He prefers to start hiking around midnight—nocturnal activity being a sensible strategy for any desert mammal—and will walk through the night and into the following afternoon without sleeping. He's sturdily built, wearing a black fleece jacket, khaki pants, sunglasses, and a broad-brimmed hat over his straight gray hair.

Thousands of tons of water surround us, sequestered in a forest of tall, green living columns; a single mature saguaro might hold as much as eight tons. Water, water, everywhere, but a lost hiker—or an illegal immigrant—would not find a drop to drink in a saguaro grove; the cactus binds its water in a viscous, slimy fluid.

The recent rain wasn't enough to save some saguaros. Paloverdes—thorny-branched trees with green photosynthetic trunks and limbs that shed their leaves during winter—are also suffering. As we wend up the flank of Wasson Peak, which rises 4,639 feet above sea level, Bertelsen's count of dead or dying plants ticks steadily upward. I ask him if we're seeing the impact of the Southwest's protracted drought.

"No question," he says. "The last time I was on this trail, maybe 10 years ago, I didn't see any dead saguaros, certainly no dead paloverdes. That's one of the reasons I wanted to come here. I want to see what's happening. That one's dying," he adds, nodding at the blackened top of a tall saguaro, multi-armed like some cactus incarnation of Vishnu. Saguaros typically don't grow arms until they're at least 75 years old. Judging by the size of its limbs, this one must have been growing for more than a century. It beat many odds during its life, starting as one of the few survivors of the tens of millions of pinhead-size black seeds produced by its parent. Like most saguaros, it probably grew in the lifesaving shade of a nurse plant—a paloverde, acacia, or ironwood tree. After 10 years it would have been just over an inch high; by 30 it would have reached two feet, its growth accelerating exponentially. It endured a drought lasting several years in the 1950s. But the current 14-year drought—the longest in at least a century—is killing it.

Saguaros may take two or three years to die. Some remain majestic even in death, standing fully upright, their gray, tubular woody skeletons flensed of all flesh. They're easy to anthropomorphize. The Tohono O'odham, one of the more than a dozen indigenous cultures of the Sonoran Desert, use the same word—*O'odham*—for both "people" and "saguaros." (*Tohono* means "desert.") In one of their old stories, the first saguaro appeared when a young girl, neglected by her mother, was transformed into a cactus, her arms forever raised to the sky.

The vagaries of life here—one saguaro dies while others on the same hillside stand replete with water—suggest to Bertelsen a biotic complexity that defies any sort of generalized explanation.

"What makes something appear and disappear? Maybe a sixteenth of an inch of rain, maybe something that is so subtle we'll never be able to figure it out," he says. "I think it's too simple to try to explain everything in terms of temperature and precipitation. Maybe rain a day earlier or later makes a big difference. There's a whole bunch of stuff that I'll see once, and then see 10 years later. I call them *desaparecidos,* the disappeared ones. Everything doesn't bloom every year. You've got to watch over a long period of time to see what's happening."

As we walk, Bertelsen keeps up a running commentary on nearly everything we pass: ocotillo, a spindly, spiny shrub with astonishing flame-red flowers; desert mistletoe, a parasite lodged in the branches of paloverdes and other trees; barrel cacti, some of which lean so far toward the sun they uproot themselves; bunches of native grasses—threeawn, bush muhly, tanglehead—that look nearly identical to me; at least two kinds of prickly pear cactus—Engelmann and mojave, their broad pads often gouged by pack rats; teddy bear cholla, singularly uncuddly, even for a cactus—five feet of "don't even think about touching me," with stubby, plump, jointed arms completely covered with barbed yellow spines that make the whole plant gleam in the bright morning light. Bertelsen carries a comb with him on his hikes in case he brushes a teddy bear—the spines will lodge in your fingers if you try to remove them from elsewhere on your body. "You know a true desert rat because they always have a comb," he says. "Not for their hair."

Bertelsen's careful observations have been honed over the years as what started as casual hikes became something more.

"I had read something by Thoreau," he says, "where he wrote how you could tell the time of month by what was blooming. So I started keeping track of blooms—not for anyone, just for curiosity. I always had a journal, but it quickly became obvious that I needed something more. So I started using a checklist. I started making comments about drought in 1994. Plants had been moving up in elevation, but it happened so gradually it was hard to see; you're too close to it. That's why stepping back and looking at the data is so useful. I was doing this because I thought it was interesting. I never thought it would be important."

"I JUST ABOUT FAINTED WHEN Dave explained what he had. There was so much information there waiting to be mined."

I'm meeting with Theresa Crimmins, an ecologist at the University of Arizona in Tucson. She's recounting how she and her husband, Michael, a climatologist at the university, first met him five years ago after a talk Michael gave on climate change.

"Dave came up to me after the talk and said, 'I have a big data set. I don't know if you would be interested,'" Theresa says. "He'd been doing this for 20 years. A paid scientist could never collect something of this magnitude."

When I mention Bertelsen's striking observation that as many as 75 percent of the desert species he's tracking don't seem to be adapting to climate change, she offers a more measured judgment. "Dave has an incredible data set," she says, "but it is segmented by mile; there are a lot more species that could be showing more subtle responses on smaller scales that we're not able to catch right now."

Michael Crimmins echoes his wife's admiration for Bertelsen's dedication. "We've worked with him to tease out the patterns in his data," he says. "A data set of, first, that quality and, second, that breadth, just doesn't exist. You would never see this in a funded project. The NSF"—the National Science Foundation—"might give you five years. You couldn't plan to collect data like this, and Dave is just so good at it, paying attention and being systematic—way better than some field scientists or grad students. Dave does it because he loves it.

"What we've found is that indeed some plants that bloomed at lower elevations when Dave began are now blooming at higher elevations," Michael continues. "But additionally, we've found a very complex dance of species. Some are responding strongly to climate change, some not so strongly. Some are blooming at a lower elevation instead of higher. The true complexity of an ecosystem is that species respond individually to climate change. They're not going to get together in a forum and decide as a biome or ecosystem that they'll do this together; they all have unique strategies to deal with climate. The response of many different species to climate change is wrapped up in Dave's data set, and it's very complex.

"We're fighting a bit of conventional wisdom here: that species will move upslope, following an envelope of perfect climate for them which is constrained by temperature. You'd expect that as it gets warmer lower down, species will move up. That works for some species, but not for all. The rate of change is the big story with climate change. When you talk with people, one of the arguments they'll throw back at you is that the climate has always changed, and that is absolutely right. It's the rate of change that is the problem right now. It's changing so quickly that it exceeds the adaptive capacity of some species."

Some adaptations are straightforward, Theresa tells me, and others less so. Warmer temperatures have allowed saguaros, for example, to expand their flowering range to higher elevations. But the flowering range of other plants—ragweed, wild carrot, and greenspot nightshade, to list just a few—has contracted, their upper-elevation limits remaining unchanged while the lower boundaries of their ranges have moved higher. "For the species showing contractions of their flowering ranges," she explains, "what we think might be going on is that warmer temperatures are becoming increasingly intolerable at the lower ends of their distributions and low-temperature triggers for signaling dormancy are not being reached."

The Crimminses published their analysis of Bertelsen's plant data last year in the journal *Global Change Biology*. They're only beginning to study his animal observations. "We need to get that into a database, primarily his information on birds," says Theresa. "He's saying, just anecdotally, that he's seeing massive declines in the number of species."

CLIMATE CHANGE IS EXACERBATING ANOTHER, more imminent threat to the Sonoran Desert: an invasive species called buffelgrass. "We have an invasion by an African grass that's capable of unhinging the Sonoran Desert," says Julio Betancourt, a paleoecologist with the United States Geological Survey. "It's more disastrous than anything climate change can throw at the desert." Betancourt, whose office is just down the hall from Theresa Crimmins, has spent most of his career studying climate change and deserts. Buffelgrass has been introduced to the southern United States as a fodder crop at various times since the late 1800s. It has now spread across southern Arizona and into Mexico, where it outcompetes many native plants. Because it evolved in a part of the world characterized by seasonal fires, it quickly reestablishes itself in areas that have been burned. Betancourt and other scientists worry that the expansion of the range of such a fast-growing species to higher altitudes as a result of global warming could convert the Sonoran Desert into a flammable savanna.

The Sonoran Desert has been essentially fire-free for 10,000 years or more. Stands of vegetation tend to be separated by wide stretches of bare, rocky ground, which limits the extent of fires. Very few Sonoran flora are adapted to fire—saguaros die after even small blazes. "Before buffelgrass was here, you could douse a paloverde with gasoline and the fire wouldn't spread," Betancourt says. In parts of the desert where buffelgrass has covered formerly bare ground, that's no longer the case. Buffelgrass and other invasive species provided much of the fuel for the fires in 2005 that burned a million acres of desert in Arizona and Nevada. The fires in Arizona, which were sparked by lightning, killed some 80 percent of the vegetation on 250,000 acres, including the largest known saguaro cactus, called the Grand One.

"That's rivaling forest fires," Betancourt says. "I think the premise of conservation will change in the Sonoran Desert. It used to be that you would try to preserve open spaces and let nature take care of itself. That's no longer the case. If you can't manage the resource, in a few decades you could have a flammable grassland instead of the saguaros and paloverdes and Gila monsters. The buffelgrass is a test. If we don't solve it, a lot of things will be moot."

"THAT'S WHERE GOD LIVES," BERTELSEN says, pointing to a peak on the southwestern horizon. We're on the summit of Wasson Peak, some 2,000 feet above the trailhead. "It's Baboquivari Peak. The Tohono O'odham say a god lives there." Tucson spreads below us; its distant edges shimmer in the warm air and seem to lap against the base of the Santa Catalina Mountains to the northeast, the site of Bertelsen's 1,270 hikes.

On the way back down the trail, Bertelsen quizzes me every 20 minutes or so, asking me to identify various grasses. By mid-afternoon I've managed only two correct answers.

"What's that?" he asks, flicking a walking pole.

Long seconds pass while I ponder a dry grass tipped with stiff bristles.

"Threeawn?"

"All right!" He seems genuinely pleased. "You've recognized three things. I do that every time I bring someone out, try to get you to recognize three things that you wouldn't have recognized before. Saguaros don't count."

Later, after pulling a few tufts of buffelgrass in a dry wash, we rest beneath a rock overhang, and the shade sharpens my appreciation of the role nurse plants play in protecting young saguaros. As we drink water, I ask Bertelsen what compels him to walk the same 10-mile trail week after week, year after year.

"I don't think I'm compulsive," he says. "I'm drawn to it. I don't feel I have to—I just really want to. Every trip there is always something different. Always."

I tell Bertelsen his words remind me of a quote from John Burroughs, the nineteenth-century American naturalist: "To find new things, take the path you took yesterday."

"Oh, I haven't heard that!" he says. "I like that. I always tell people, it's not 1,270 hikes. It's a hike of 12,700 miles. It's one journey. It's not a separate thing. That's exactly the way I feel about it. It's one long, continuous walk."

BURKHARD BILGER

Nature's Spoils

FROM *THE NEW YORKER*

> *Opportunivores. Techno-peasants. Drinkers of raw milk and eaters of "high meat." Burkhard Bilger encounters these—and more—in search of the primal diet.*

THE HOUSE AT 40 CONGRESS STREET WOULDN'T have been my first choice for lunch. It sat on a weedy lot in a dishevelled section of Asheville, North Carolina. Abandoned by its previous owners, condemned by the city, and minimally rehabilitated, it was occupied—perhaps infested is a better word—by a loose affiliation of opportunivores. The walls and ceilings, chicken coop and solar oven were held together with scrap lumber and drywall. The sinks, disconnected from the sewer, spilled their effluent into plastic buckets, providing water for root crops in the gardens.

The whole compound was painted a sickly greenish gray—the unhappy marriage of twenty-three cans of surplus paint from Home Depot. "We didn't put in the pinks," Clover told me.

Clover's pseudonym both signalled his emancipation from a wasteful society and offered a thin buffer against its authorities. "It came out of the security culture of the old Earth First! days," another opportunivore told me. "If the Man comes around, you can't give him any incriminating information." Mostly, though, the names fit the faces: Clover was pale, slender, and sweet-natured, with fine blond hair gathered in a bun. His neighbor Catfish had droopy whiskers and fleshy cheeks. There were four men and three women in all, aged twenty to thirty-five, crammed into seven small bedrooms. Only one had a full-time job, and more than half received food stamps. They relied mostly on secondhand bicycles for transportation, and each paid two hundred dollars or less in rent. "We're just living way simple," Clover said. "Super low-impact, deep green."

Along one wall of the kitchen, rows of pine and wire shelves were crowded with dumpster discoveries, most of them pristine: boxes of organic tea and artisanal pasta, garlic from Food Lion, baby spring mix from Earth Fare, tomatoes from the farmers' market. About half the household's food had been left somewhere to rot, Clover said, and there was often enough to share with Asheville's other opportunivores. (A couple of months earlier, they'd unearthed a few dozen cartons of organic ice cream; before that, enough Odwalla juices to fill the bathtub.) Leftovers were pickled or composted, brewed into mead or, if they looked too dicey, fed to the chickens. "We have our standards," a young punk with a buzz-cut scalp and a skinny ponytail told me. "We won't dumpster McDonald's." But he had eaten a good deal of scavenged sushi, he said—it was all right, as long as it didn't sit in the dumpster overnight—and his housemate had once scored a haggis. "Oh no, no," she said, when I asked if she'd eaten it. "It was canned."

Lunch that day was lentil soup, a bowl of which was slowly con-

gealing on the table in front of me. The carrots and onions in it had
come from a dumpster behind Amazing Savings, as had the lentils,
potatoes, and most of the spices. Their color reminded me a little of
the paint on the house. Next to me, Sandor Katz scooped a spoonful
into his mouth and declared it excellent. A self-avowed "fermentation fetishist," Katz travels around the country giving lectures and
demonstrations, spreading the gospel of sauerkraut, dill pickles, and
all foods transformed and ennobled by bacteria. His two books—
Wild Fermentation and *The Revolution Will Not Be Microwaved*—
have become manifestos and how-to manuals for a generation of
underground food activists, and he's at work on a third, definitive
volume. Lunch with the opportunivores was his idea.

Katz and I were on our way to the Green Path, a gathering of
herbalists, foragers, raw-milk drinkers, and roadkill eaters in the
foothills of the Smoky Mountains. The groups in Katz's network
have no single agenda or ideology. Some identify themselves as
punks, others as hippies, others as evangelical Christians; some live
as rustically as homesteaders—the "techno-peasantry," they call
themselves; others are thoroughly plugged in. If they have a connecting thread, it's their distrust of "dead, anonymous, industrial-
ized, genetically engineered, and chemicalized corporate food," as
Katz has written. Americans are killing themselves with cleanliness,
he believes. Every year, we waste forty per cent of the food we produce, and process, pasteurize, or irradiate much of the rest, steriliz-
ing the live cultures that keep us healthy. Lunch from a Dumpster
isn't just a form of conservation; it's a kind of inoculation.

"This is a modern version of the ancient tradition of gleaning,"
Katz said. "When the harvest is over, the community has a common-
law right to pick over what's left." I poked at the soup with my spoon.
The carrots seemed a little soft—whether overcooked or overripe, I
couldn't tell—but they tasted all right. I asked the kid with the ponytail if he'd ever brought home food that was spoiled. "Oh, hell
yes!" he said, choking back a laugh. "Jesus Christ, yes!" Then he

shrugged, suddenly serious. "It happens: diarrhea, food poisoning. But I think we've developed pretty good immune systems by now."

To most cooks, a kitchen is a kind of battle zone—a stain-less-steel arena devoted to the systematic destruction of bacteria. We fry them in oil and roast them in ovens, steam them, boil them, and sluice them with detergents. Our bodies are delicate things, easily infected, our mothers taught us, and the agents of microscopic villainy are everywhere. They lurk in raw meat, raw vegetables, and the yolks of raw eggs, on the unwashed hand and in the unmuffled sneeze, on the grimy countertop and in the undercooked pork chop.

Or maybe not. Modern hygiene has prevented countless colds, fevers, and other ailments, but its central premise is hopelessly outdated. The human body isn't besieged; it's saturated, infused with microbial life at every level. "There is no such thing as an individual," Lynn Margulis, a biologist at the University of Massachusetts at Amherst, told me recently. "What we see as animals are partly just integrated sets of bacteria." Nearly all the DNA in our bodies belongs to microorganisms: they outnumber our own cells nine to one. They process the nutrients in our guts, produce the chemicals that trigger sleep, ferment the sweat on our skin and the glucose in our muscles. ("Humans didn't invent fermentation," Katz likes to say. "Fermentation created us.") They work with the immune system to mediate chemical reactions and drive out the most common infections. Even our own cells are kept alive by mitochondria—the tiny microbial engines in their cytoplasm. Bacteria are us.

"Microbes are the minimal units, the basic building blocks of life on earth," Margulis said. About half a billion years ago, land vertebrates began to encase themselves in skin and their embryos in protective membranes, sealing off the microbes inside them and fostering ever more intimate relations with them. Humans are the acme of that evolution—walking, talking microbial vats. By now, the communities we host are so varied and interdependent that it's hard to tell friend from enemy—the bacteria we can't live with from those we can't live without. E. coli, Staphylococcus aureus, and the bacteria responsible for meningitis and stomach ulcers all live peaceably inside us most of the time, turning dangerous only on rare occasions and for reasons that are poorly understood. "This cliché nonsense about good and bad bacteria, it's so insidious," Margulis said. "It's this Western, dichotomized, Cartesian thing. . . . Like Jesus rising."

In the past decade, biologists have embarked on what they call the second human-genome project, aimed at identifying every bacterium associated with people. More than a thousand species have been found so far in our skin, stomach, mouth, guts, and other body parts. Of those, only fifty or so are known to harm us, and they have been studied obsessively for more than a century. The rest are mostly new to science. "At this juncture, biologists cannot be blamed for finding themselves in a kind of 'future shock,'" Margaret McFall-Ngai, an expert in symbiosis at the University of Wisconsin at Madison, wrote in *Nature Reviews Microbiology* two years ago. Or, as she put it in an earlier essay, "We have been looking at bacteria through the wrong end of the telescope."

Given how little we know about our inner ecology, carpet-bombing it might not always be the best idea. "I would put it very bluntly," Margulis told me. "When you advocate your soaps that say they kill all harmful bacteria, you are committing suicide." The bacteria in the gut can take up to four years to recover from a round of antibiotics, recent studies have found, and the steady assault of detergents, preservatives, chlorine, and other chemicals also takes its toll. The immune system builds up fewer antibodies in a sterile environment; the deadliest pathogens can grow more resistant to antibiotics; and innocent bystanders such as peanuts or gluten are more likely to provoke allergic reactions. All of which may explain why a number of studies have found that children raised on farms are less susceptible to allergies, asthma, and autoimmune diseases. The cleaner we are, it sometimes seems, the sicker we get.

"We are living in this cultural project that's rarely talked about," Katz says. "We hear about the war on terror. We hear about the war on drugs. But the war on bacteria is much older, and we've all been indoctrinated into it. We have to let go of the idea that they're our enemies." Eating bacteria is one of life's great pleasures, Katz says. Beer, wine, cheese, bread, cured meats, coffee, chocolate: our best-loved foods are almost all fermented. They start out bitter, bland, cloying, or indigestible and are remade by microbes into something magnificent.

Fermentation is a biochemical magic trick—a benign form of rot. It's best known as the process by which yeast turns sugar into alcohol, but an array of other microorganisms and foods can ferment as well. In some fish dishes, for instance, the resident bacteria digest amino acids and spit out ammonia, which acts as a preservative. Strictly speaking, all fermentation is anaerobic (it doesn't consume oxygen); most rot is aerobic. But the two are separated less by process than by product. One makes food healthy and delicious; the other not so much.

Making peace with microorganisms has its risks, of course. E. coli can kill you. Listeria can kill you. Basic hygiene and antibiotic over-kill aren't hard to tell apart at home, but the margin of error shrinks dramatically in a factory. Less than a gram of the bacterial toxin that causes botulism, released into the American milk supply, could poison a hundred thousand people, the National Academy of Sciences estimated in 2005. And recent deaths and illnesses from contaminated beef, spinach, and eggs have persuaded food regulators to clamp down even harder. While Katz's followers embrace their bacterial selves, the Obama Administration has urged Congress to pass a comprehensive new set of food-safety laws, setting the stage for a culture war of an unusually literal sort. "This is a revolution of the everyday," Katz says, "and it's already happening."

* * *

WHEN KATZ PICKED ME UP in Knoxville at the beginning of our road trip, the back seat of his rented Kia was stacked with swing-top bottles and oversized Mason jars. They were filled with foamy, semi-opaque fluids and shredded vegetables that had been fermenting in his kitchen for weeks. A sour, pleasantly funky aroma pervaded the cabin, masking the new-car smell of industrial cleaners and off-gassing plastics. It was like driving around in a pickle barrel.

Physically speaking, food activists tend to present a self-negating argument. The more they insist on healthy eating the unhealthier they look. The pickier they are about food the more they look like they could use a double cheeseburger. Katz was an exception. At forty-eight, he had clear blue eyes, a tightly wound frame, and ropy forearms. His hands were calloused and his skin was ruddy from hours spent weeding his commune's vegetable patches and herding its goats. He wore his hair in a stubby Mohawk, his beard in bushy muttonchops. If not for his multiple earrings and up-to-the-minute scientific arguments, he might have seemed like a figure out of the nineteenth century, selling tonics and bromides from a painted wagon.

Katz was a political activist long before he was a fermentation fetishist. Growing up on New York's Upper West Side, the eldest son of progressive Polish and Lithuanian Jews, he was always involved in one campaign or another. At the age of ten, in 1972, he spent his afternoons on street corners handing out buttons for George McGovern. At eleven, he was a campaign volunteer for the mayoral candidate Al Blumenthal. When he reached sixth grade and found that one of the city's premier programs for gifted students, Hunter College High School, was only for girls, he helped bring an anti-discrimination suit that forced it to turn coed. He later served on the student council with Elena Kagan, the future Supreme Court Justice. "The staggered lunch hour was our big issue," he says.

At Brown, as an undergraduate, Katz became a well-known

figure: a bearish hippie in the Abbie Hoffman mold, with a huge head of curly hair. His causes were standard issue for the time: gay rights, divestment from South Africa, U.S. out of Central America (as a senior, he and a group of fellow-activists placed a C.I.A. recruiter under citizen's arrest). Yet Katz lacked the usual stridency of the campus radical. "I remember a particular conversation in 1982 or '83," his classmate Alicia Svigals, who went on to found the band the Klezmatics, told me. "We were standing on a street corner in Providence, and I said, 'Sandy, I think I might be a lesbian.' And he said, 'Oh, I think I might be gay.' At the time, that was a huge piece of news. It wasn't something you said lightly. But his reaction was 'How wonderful and exciting! How fantastic! This is going to be so much fun!' The world was about to be made new—and so easily."

After graduation, Katz moved back to New York. He took a job as the executive director of Westpride, a lobbying group opposed to a massive development project on the Upper West Side. (The developer, Donald Trump, was eventually forced to scale down his plans.) As the AIDS epidemic escalated, in the late eighties, Katz became an organizer for ACT UP and a columnist for the magazine *OutWeek*. His efforts on both fronts caught the eye of Ruth Messinger, the Manhattan borough president, who hired him in 1989 as a land-use planner and as a de-facto liaison to the gay community. "He was a spectacular person," Messinger told me recently. "Creative and flamboyant and fun to be around. He just had a natural instinct and talent as an organizer of people." Messinger was thinking of running for mayor (she won the Democratic nomination in 1997, only to get trounced by Rudolph Giuliani), and Katz's ambitions rose with hers. "I would fantasize about what city agency I wanted to administer," he recalls. Then, in 1991, he found out he was H.I.V.-positive.

Katz had never been particularly promiscuous. He'd had his first gay sexual experience at the age of twenty-one, crossing the country on a Green Tortoise bus, and had returned to New York just as its bathhouse days were waning. He'd never taken intravenous drugs and had avoided the riskiest sexual activities. The previous H.I.V. tests he'd taken had come back inconclusive—perhaps, he reasoned, because of a malarial infection that he'd picked up in West Africa. "I have no idea how it happened," he told me. "I remember walking out of the doctor's office in such a daze. I was just utterly shell-shocked."

The virus wasn't necessarily a death sentence, though an effective treatment was years away. But it did transform Katz's political ambitions. "They just dematerialized," he told me. For all his iconoclasm, he had always dreamed of being a United States senator. Now he focussed on curing himself. He cut back his hours and moved from his parents' apartment to the East Village. So many of his friends had died while on AZT and other experimental drugs that he decided to search for alternatives. He had already taken up yoga and switched to a macrobiotic diet. Now he began to consult with herbalists, drink nettle tea, and wander around Central Park gathering medicinal plants. "I got skinny, skinny, skinny," he says. "My friends thought I was wasting away."

New York's relentless energy had always helped drive his ambitions, but now he found that it wore him out. About a year after his diagnosis, Katz went to visit some friends in New Orleans who had rented a crash pad for Mardi Gras. Among the characters there, he met a man from a place near Nashville called Hickory Knoll (I've changed the name at Katz's request). Founded in the early seventies by a group of back-to-the-landers, Hickory Knoll was something of a legend in the gay community: a queer sanctuary in the heart of the Bible Belt. "I was a typical New Yorker," Katz says. "I considered the idea of living in Tennessee absurd." Still, he was intrigued. Hickory Knoll had no television or hot running water—just goats, vegetable gardens, and gay men. Maybe it was just what he needed.

HICKORY KNOLL LIES JUST UP the road from a Bible camp, in an airy forest of tulip poplar and dogwood, maple, mountain laurel, pawpaw, and persimmon. The camp and the commune share a hill-top, a telephone cable, and, if nothing else, a belief in spiritual re-newal: "Want a new life?" a sign in front of a local church asked as I drove past. "God accepts trade-ins." When Katz first arrived, in the spring of 1992, the paulownia trees were in bloom, scattering the ground with lavender petals. As he walked down the gravel path, the forest canopy opened up and a cabin of hand-hewn chestnut logs, built in the eighteen-thirties, appeared in the sunlight below. "It was a beautiful arrival," he says.

The commune had a shifting cast of about fifteen members, some of whom had lived there for decades. It billed itself as a radical faerie sanctuary, though the term was notoriously slippery—the faerie movement, begun in the late seventies by gay-rights activists, em-braced everyone from transvestites to pagans and anarchists, their common interest being a focus on nature and spirituality. Street kids from San Francisco, nudists from Nashville, a Mexican minister coming out of the closet: all found their way, somehow, to central Tennessee. Most were gay men, though anyone was welcome, and the great majority had never lived on the land before. "Sissies in the wood," one writer called it, after tussling over camping arrange-ments with a drag queen in four-inch heels.

New arrivals stayed in the cabin "downtown," which had been fitfully expanded to encompass a library, living room, dining room, and kitchen, with four bedrooms upstairs. Farther down the path were a swaybacked red barn, a communal shower, a pair of enor-mous onion-domed cisterns, and a four-seater outhouse. The charge for room and board was on a sliding scale starting at seventy-five dollars a month, with a tacit agreement, laxly enforced, to pitch in—milking goats, mending fences, or just greeting new arrivals. Those who stayed eventually built houses along the ridges or bought adja-cent land and started homesteads and communes of their own. In the spring, at the annual May Day celebration, their numbers grew to several hundred. "The gayborhood just keeps on growing,"

Weeder, Hickory Knoll's oldest member, told me one evening as we were sitting on the front porch of the cabin. "We're a pretty good voting bloc."

Inside, half a dozen men were preparing dinner. Food is the great marker of the day at Hickory Knoll—the singular goal toward which most labor and creativity tend. On my visit, the kitchen seemed to be staffed by at least three cooks at all times, cutting biscuits, baking vegan meat loaf, washing kale; one of them, a gangly Oklahoman named Lady Now, worked in the nude. "Real estate determines culture," Katz likes to say, and the maxim is doubly true among underground food movements. Urban squatters gravitate toward freeganism and dumpster diving, homesteaders toward raw milk and roadkill. At Hickory Knoll, the slow pace, lush gardens, and communal isolation are natural incubators for fermented food, though Katz didn't realize it right away. "It took a while for the New York City to wear off," Weeder told me. "Overanalyzing everything. Where am I going to go tonight? There really is nowhere to go."

That first year, a visitor named Crazy Owl brought some miso as barter for his stay, inspiring Katz to make some of his own. Miso, like many Asian staples, is usually made of fermented soybeans. The beans are hard to stomach alone, no matter how long they're cooked. But once inoculated with koji—the spores of the Aspergillus oryzae mold—they become silken and delicious. The enzymes in the mold predigest the beans, turning starches into sugars, breaking proteins into amino acids, unlocking nutrients from leaden compounds. A lowly bean becomes one of the world's great foods.

Katz experimented with more and more fermented dishes after that. He made tempeh, natto, kombucha, and kefir. He recruited friends to chew corn for chicha—an Andean beer brewed with the enzymes in human saliva. At the Vanderbilt library, in Nashville, he worked his way through the "Handbook of Indigenous Fermented Foods" (1983), by the Cornell microbiologist Keith Steinkraus. When he'd gathered a few dozen recipes, he printed a pamphlet and

sold some copies to a bookstore in Maine and a permaculture magazine in North Carolina. The pamphlet led to a contract from a publisher, Chelsea Green, and the release of "Wild Fermentation," in
2003. The book was only a modest success at first, but sold more
copies each year—some seventy thousand in all. Soon Katz was
crisscrossing the country in his car, shredding cabbage in the aisles
of Whole Foods or Trader Joe's, preaching the glories of sauerkraut.

"FERMENTED FOODS AREN'T CULINARY NOVELTIES," he
told me one morning. "They aren't cupcakes. They're a major survival food." We were standing in his test kitchen, in the basement of
a farmhouse a few miles down the road from Hickory Knoll. Katz
had rented the space two years earlier, when his classes and cooking
projects outgrew the commune's kitchen, and outfitted it with secondhand equipment: a triple sink, a six-burner stove, a freezer, and
two refrigerators, one of them retrofitted as a tempeh incubator.
Along one wall, a friend had painted a psychedelic mural showing a
man conversing with a bacterium. Along another, Katz had pinned a
canticle to wild fermentation, written by a Benedictine nun in New
York. A haunch of venison hung in back, curing for prosciutto, surrounded by mismatched jars of sourdough, goat kefir, sweet potato
fly, and other ferments, all bubbling and straining at their lids. "It's
like having pets," Katz said.

The kitchen had the same aroma as Katz's car, only a few orders of
magnitude funkier: the smell of life before cold storage. "We are
living in the historical bubble of refrigeration," Katz said, pulling a
jar of bright-pink-and-orange sauerkraut off the shelf. "Most of these
food movements aren't revolutionary so much as conservative. They
want to bring back the way food has been."

Fermentation, like cooking with fire, is one of the initial conditions of civilization. The alcohol and acids it produces can preserve
fruits and grains for months and even years, making sedentary soci-

ety possible. The first ferments happened by accident—honey water turned to mead, grapes to vinegar—but people soon learned to recreate them. By 5400 B.C., the ancient Iranians were making wine. By 1800 B.C., the Sumerians were worshipping Ninkasi, the goddess of beer. By the first century B.C., the Chinese were making a precursor to soy sauce.

Katz calls fermentation the path of least resistance. "It's what happens when you do nothing," he says. Or, rather, if you do one or two simple things. A head of cabbage left on a counter will never turn to sauerkraut, no matter how long it sits there. Yeasts, molds, and a host of bacteria will attack it, digesting the leaves till all that's left is a puddle of black slime. To ferment, most food has to be protected from the air. It can be sealed in a barrel, stuffed in a casing, soaked in brine, or submerged in its own juices—anything, as long as oxygen doesn't touch it. The sauerkraut Katz was holding had been made ten days earlier. I'd watched him shred the cabbage—one head of red and one of green—sprinkle it with two tablespoons of salt to draw out the water, and throw in a few grated carrots. He'd scrunched everything together with his hands, to help release the juice, and packed it in a jar until the liquid rose to the top. "I would suggest not sealing it too tightly," he said, as he clamped down the lid. "Some jars will explode."

Three waves of bacteria had colonized the kraut since then, each one changing the chemical environment just enough to attract and fall victim to the next—like yuppie remodellers priced out of their own neighborhood. Sugars had been converted to acids, carbon dioxide, and alcohol. Some new nutrients had been created: B vitamins, for instance, and isothiocyanates, which laboratory studies have found to inhibit lung, liver, breast, and other cancers. Other nutrients were preserved, notably Vitamin C. When Captain Cook circled the globe between 1772 and 1775, he took along thirty thousand pounds of sauerkraut, and none of his crew died of scurvy.

I tried a forkful from Katz's jar, along with a slab of his black-rice tempeh. The kraut was crunchy and tart—milder than any I'd had from a store and much fresher tasting. "You could eat it after two weeks, you could eat it after two months, and if you lived in a cold environment and had a root cellar you could eat it after two years," Katz said. The longer it fermented the stronger it would get. His six-month-old kraut, made with radishes and Asian greens, was meaty, pungent, and as tender as pasta—the enzymes in it had broken down the pectin in its cell walls. Some people like it that way, he said. "When this Austrian woman tasted my six-week-old sauerkraut, she said, 'That's O.K.—for coleslaw.'"

While we were eating, the front door banged open and a young man walked in carrying some baskets of fresh-picked strawberries. He had long blond hair and hands stained red with juice. His name was Jimmy, he said. He lived at Hickory Knoll but was doing some farming up the road. "We originally grew herbs and flowers and planted them in patterns," he said. "But people were like, 'What are those patterns you're makin'? They don't look Christian to me.'" The locals were usually pretty tolerant, Katz said. In eighteen years, the worst incidents that he could recall were a few slashed tires and some teen-agers yelling "Faggots!" from the road and shooting shot-guns in the air. Rural Tennessee is a "don't ask, don't tell" sort of place, where privacy is the one inalienable right. But Jimmy's fancy crop might have counted as a public display. He laughed and handed me a berry, still warm from the sun. "They're not only organic," he said. "They're grown with gay love."

The fruit was sugar-sweet and extravagantly fragrant—a distillation of spring. But the sauerkraut was the more trustworthy food. An unwashed fruit or vegetable may host as many as a million bacteria per gram, Fred Breidt, a microbiologist with the United States Department of Agriculture and a professor of food science at North Carolina State University, told me. "We've all seen the cases," Katz said. "The runoff from agriculture gets onto a vegetable, or there's fecal matter from someone who handled it. Healthy people will get diarrhea; an elderly person or a baby might get killed. That's a possibility with raw food." If the same produce were fermented, its native bacteria would drive off the pathogens, and the acids and alcohol they produce would prevent any further infection. Breidt has yet to find a single documented case of someone getting sick from contaminated sauerkraut. "It's the safest food there is," Katz said.

SAUERKRAUT IS KATZ'S GATEWAY DRUG. He lures in novices with its simplicity and safety, then encourages them to experiment with livelier cultures, more offbeat practices. *The Revolution Will Not Be Microwaved* moves from anodyne topics such as seed saving and urban gardening to dirt eating, feral foraging, cannabis cookery, and the raw-milk underground. Unlike many food activists, Katz has a clear respect for peer-reviewed science and he prefaces each discussion with the appropriate caveats. Yet his message is clear: "Our food system desperately demands subversion," he writes. "The more we sterilize our food to eliminate all theoretical risk, the more we diminish its nutritional quality."

On the first day of our road trip, not long after our lunch with the opportunivores, Katz and I paid a visit to a man he called one of the kingpins of underground food in North Carolina. Garth, as I'll call him, was a pale, reedy figure in his fifties with wide, spectral eyes. His linen shirt and suspenders hung slackly on his frame, and his sunken cheeks gave him the look of a hardscrabble farmer from a century ago. "I was sick for seventeen years," he told us. "Black circles under my eyes, weighed less than a hundred pounds. It didn't seem like I'd get very far." Doctors said that he had severe chemical sensitivities and a host of ailments—osteoporosis, emphysema, edema, poor circulation—but they seemed incapable of curing him. He tried veganism for a while, but only got weaker. "It's just not a good diet for skinny people," he said. So he went to the opposite extreme.

Inside his bright country kitchen, Garth carefully poured us each a glass of unpasteurized goat milk, as if proffering a magic elixir. The milk was pure white and thick as cream. It had a long, flowery bloom and a faint tanginess. Raw milk doesn't spoil like pasteurized milk. Its native bacteria, left to multiply at room temperature, sour it into something like yogurt or buttermilk, only much richer in cultures. It was the mainstay of Garth's diet, along with raw butter, cream, and daily portions of raw liver, fish, chicken, or beef. He was still anything but robust, but he had enough energy to work long hours in the garden for the first time in years. "It enabled me to function," he said.

Raw milk brings the bacterial debate down to brass tacks. Drinking it could be good for you. Then again it could kill you. Just where the line between risk and benefit lies is a matter of fierce dispute—not to mention arrests, lawsuits, property seizures, and protest marches. In May, for instance, raw-milk activists, hoping to draw attention to a recent crackdown by Massachusetts agricultural authorities, milked a Jersey cow on Boston Common and staged a drink-in.

Retail sales of raw milk are illegal in most states, including North Carolina, but people drink it anyway. Some dairy owners label the milk for pet consumption only (though at two to five times the cost of pasteurized, it's too rich for most cats). Others sell it at farm stands or through herd-share programs. In my neighborhood in Brooklyn, the raw-milk coöperative meets every month in the aisles of a gourmet deli. The milk is trucked in from Pennsylvania—a violation of federal law, which prohibits the interstate transport of raw milk—but no one seems to mind. Garth buys milk from a local farmer and sells it out of his house. "It's illegal," he told me. "But it gets to the point where living is illegal."

The nutritional evidence both for and against raw milk is somewhat sketchy; much of it dates from before the Second World War, when raw milk was still legal. The Food and Drug Administration, in a fact sheet titled "The Dangers of Raw Milk," insists that pasteurization "DOES NOT reduce milk's nutritional value." The temperature of the process, well below the boiling point, is meant to kill pathogens and leave nutrients intact. Yet raw-milk advocacy groups, such as the Weston A. Price Foundation, in Washington, D.C., point to a number of studies that suggest the opposite. An array of vitamins, enzymes, and other nutrients are destroyed, diminished, or denatured by heat, they say. Lactase, for instance, is an enzyme that breaks down lactose into simpler sugars that the body can better digest. Raw milk often contains lactobacilli and bifidobacteria that produce lactase, but neither the bacteria nor the enzyme can survive pasteurization. In one survey of raw-milk drinkers in Michigan and Illinois, eighty-two per cent of those who had been diagnosed as lactose intolerant could drink raw milk without digestive problems. (A more extreme view, held by yet another dietary faction, is that people shouldn't be drinking milk at all—that it's a food specifically designed for newborns of other species, and as such inimical to humans.)

To the F.D.A., the real problem with milk isn't indigestion but contamination. Poor hygiene and industrial production are a toxic combination. One sick cow, one slovenly worker can contaminate the milk of a dozen dairies. In 1938, a quarter of all disease outbreaks from contaminated food came from milk, which had been known to carry typhoid, tuberculosis, diphtheria, and a host of other diseases. More recently, between 1998 and 2008, raw milk was responsible for eighty-five disease outbreaks in more than twenty states, including more than sixteen hundred illnesses, nearly two hundred hospitalizations, and two deaths. "Raw milk is inherently dangerous," the F.D.A. concludes. "It should not be consumed by anyone at any time for any purpose."

Thanks in large part to pasteurization, dairy products now account for less than five per cent of the foodborne disease outbreaks in America every year. Smoked seafood is six times more likely than pasteurized milk to contain listeria; hot dogs are sixty-five times more likely, and deli meats seventy-seven times more likely. "Every now and then, I meet people in the raw-milk movement who say, 'We have to end pasteurization now!'" Katz told me. "We can't end pasteurization. It would be the biggest disaster in the world. There would be a lot of dead children around."

Still, he says, eating food will always entail a modicum of risk. In an average year, there are seventy-six million cases of food poisoning in America, according to the Centers for Disease Control. Raw milk may be more susceptible to contamination than most foods (though it's still ten times less likely to contain listeria than deli meat is). But just because it can't be produced industrially doesn't mean it can't be produced safely, in smaller quantities. Wisconsin has some thirteen thousand dairies, about half of which, local experts estimate, are owned by farmers who drink their own raw milk. Yet relatively few people have been known to get sick from it. "If this were such a terrible cause and effect, we would be in the newspaper constantly," Scott Rankin, the chairman of the food-science department at the University of Wisconsin at Madison, and a member of the state's raw-milk working group, told me. "Clearly there is an argument to be made in the realm of, yeah, this is a tiny risk."

The country's largest raw-milk dairy is Organic Pastures, in Fresno, California. Its products are sold in three hundred and seventy-five stores and serve fifty thousand people a week. "Nobody's dying," the founder and C.E.O., Mark McAfee, told me. In ten years, only two of McAfee's customers have reported serious food poisoning, he says, and none of the bacteria in those cases could be traced to his dairy. Raw milk is rigorously tested in California and has to meet strict limits for bacterial count. The state's standards hark back to the early days of pasteurization, when many doctors considered raw milk far more nutritious than pasteurized, and separate regulations insured its cleanliness. Dealing with live cultures, Katz and McAfee argue, forces dairies to do what all of agriculture should be doing anyway: downsize, localize, clean up production. "We need to go back a hundred and fifty years," McAfee told me. "Going back is what's going to help us go forward."

A CENTURY AND A HALF is an eternity in public-health terms, but to followers of the so-called primal diet it's not nearly long enough. Humans have grown suicidally dainty, many of them say, and even a diet enriched by fermented foods and raw milk is too cultivated by half. Our ancestors were rough beasts: hunters, gatherers, scavengers, and carrion eaters, built to digest any rude meal they could find. Fruits and vegetables were a rarity, grains nonexistent. The human gut was a wild kingdom in those days, continually colonized and purged by parasites, viruses, and other microorganisms picked up from raw meat and from foraging. What didn't kill us, as they say, made us stronger.

A few miles north of downtown Asheville, in a small white farmhouse surrounded by trees, two of Katz's acquaintances were doing their best to emulate early man. Steve Torma ate mostly raw meat and raw dairy. His partner, Alan Muskat, liked to supplement his diet with whatever he could find in the woods: acorns, puffballs, cicadas and carpenter ants, sumac leaves and gypsy-moth caterpillars. Muskat was an experienced mushroom hunter who had provisioned a number of restaurants in Asheville, and much of what he served us was surprisingly good. The ants, collected from his woodpile in the winter when they were too sluggish to get away, had a snappy texture and bright, tart flavor—like organic Pop Rocks. (They were full of formic acid, which gets its name from the Latin word for ant.) He brought us a little dish of toasted acorns, cups of honey-sweetened sumac tea, and goblets of a musky black broth made from decomposed inky-cap mushrooms. I felt, for a moment, as if I'd stumbled upon a child's tea party in the woods.

The primal diet has found a sizable following in recent years, particularly in Southern California and, for some reason, Chicago. Its founder, Aajonus Vonderplanitz, a sixty-three-year-old former soap-opera actor and self-styled nutritionist, claims that it cured him of autism, angina, dyslexia, juvenile diabetes, multiple myeloma, and stomach cancer, as well as psoriasis, bursitis, osteoporosis, tooth decay, and "mania created by excessive fruit." Vonderplanitz recommends eating roughly eighty-five per cent animal products by volume, supplemented by no more than one fruit a day and a pint or so of "green drink"—a purée of fruits and vegetable juices. (Whole vegetable fibres, he believes, are largely indigestible.) The diet's most potent component, though, is an occasional serving of what Vonderplanitz calls "high meat."

Torma ducked into the back of the house and returned with a swing-top jar in his hands. Inside lay a piece of organic beef, badly spoiled. It was afloat in an ochre-colored puddle of its own decay, the muscle and slime indistinguishable, like a slug. High meat is the flesh of any animal that has been allowed to decompose. Torma keeps his portions sealed for up to several weeks before ingesting them, airing them out every few days. (Like the bacteria in sauerkraut, those which cause botulism are anaerobic; fermentation destroys them, but they sometimes survive in sealed meats—botulus, in Latin, means sausage.) Vonderplanitz says that he got high meat and its name from the Eskimos, who savor rotten caribou and seal. A regular serving of decayed heart or liver can have a "tremendous Viagra effect" on the elderly, Vonderplanitz told me recently. The first few bites, though, can be rough going. "I still have some resistance to it," Torma admitted. "But the health benefits! I'm fifty-two now. I started this when I was forty-two, and I feel like I'm in my twenties."

Primal eating has its detractors: *The Times* of London recently dubbed it "the silliest diet ever." Most of us find whole vegetables perfectly digestible. The notion that parasites and viruses are good for us would be news to most doctors. And even Vonderplanitz and his followers admit that high meat sometimes leaves them ill and explosively incontinent. They call it detoxification.

Still, radical measures like these have had some surprising successes. In a case published last year in the *Journal of Clinical Gastro-enterology*, a sixty-one-year-old woman was given an entirely new set of intestinal bacteria. The patient was suffering from severe diarrhea, Janet Jansson, a microbial ecologist at Lawrence Berkeley National Laboratory, told me. "She lost twenty-seven kilograms and was confined to a diaper and a wheelchair." To repopulate her colon with healthy microbes, Jansson and her collaborators arranged for a fecal transplant from the patient's husband. "They just put it in a Waring blender and turned it into a suppository," Jansson says. "It sounds disgusting, but it cured her. When we got another sample from her, two days later, she had adopted his microbial community." By then, her diarrhea had disappeared.

Other experiments have been even more dramatic. At Washington University, in St. Louis, the biologist Jeff Gordon has found that bacteria can help determine body weight. In a study published in *Nature* in 2006, Gordon and his lab-mates, led by Peter Turnbaugh, took a group of germ-free mice, raised in perfect sterility, and divided it in two. One group was inoculated with bacteria from normal mice; the other with bacteria from mice that had been bred to be obese. Both groups gained weight after the inoculation, but those with bacteria from obese mice had nearly twice the percentage of body fat by the end of the experiment. Later, the same lab took normal mice, fed them until they were fat, and transplanted their bacteria into other normal mice. Those mice grew fat, too, and the same pattern held true when the mice were given bacteria from obese people. "It's a positive-feedback loop," Ruth Ley, a biologist from Gordon's lab who now teaches at Cornell, told me. "Whether you're genetically obese or obese from a high-fat diet, you end up with a microbial community that is particularly good at extracting calories. It could mean that an obese person can extract an extra five or ten calories out of a bowl of Cheerios."

Biologists no longer doubt the depth of our dependence on bacteria. Jansson avoids antibiotics unless they're the only option, and eats probiotic foods like yogurt and prebiotic foods like yacón, a South American root that nourishes bacteria in the gut. Until we understand more about this symbiosis, she and others say, it's best to ingest the cultures we know and trust. "What's beautiful about fermenting vegetables is that they're naturally populated by lactic-acid bacteria," Katz said. "Raw flesh is sterile. You're just culturing whatever was on the knife."

When Torma unclamped his jar, a sickly-sweet miasma filled the air—an odor as natural as it was repellent. Decaying meat produces its own peculiar scent molecules, I later learned, with names like putrescine and cadaverine. I could still smell them on my clothes hours later. Torma stuck two fingers down the jar and fished out a long, wet sliver. "Want a taste?" he said.

It was the end of a long day. I'd spent most of it consuming everything set before me: ants, acorns, raw milk, Dumpster stew, and seven kinds of mead, among other delicacies. But even Katz took a pass on high meat. While Torma threw back his head and dropped in his portion, like a seal swallowing a mackerel, we quietly took our leave. "You have to trust your senses," Katz said, as we were driving away. "To me, that smelled like death."

KATZ HAS LIVED WITH H.I.V. for almost two decades. For many years, he medicated himself with his own ferments and local herbs—chickweed, yellow dock, violet leaf, burdock root. But periodic tests at the AIDS clinic in Nashville showed that his T-cell count was still low. Then, in the late nineties, he began to lose weight. He often felt listless and mildly nauseated. At first, he assumed that he was just depressed, but the symptoms got worse. "I started feeling light-headed a lot and I had a couple of fainting episodes," he told me. "It dawned on me very slowly that I was suffering from classic AIDS wasting syndrome."

By then, an effective cocktail of AIDS drugs had been available for almost three years. Katz had seen it save the life of one of his neighbors in Tennessee. "It was a really dramatic turnaround," he told me. "But I didn't want my life to be medically managed. I had a real reluctance to get on that treadmill." In the summer of 1999, he took a road trip to Maine to visit friends, hoping to snap out of his funk. By the time he got there, he was so exhausted that he couldn't get up for days. "I remember what really freaked me out was trying to balance my checkbook," he says. "I couldn't even do simple subtraction. It was like my brain wasn't functioning anymore." He had reached the end of his alternatives.

Katz doesn't doubt that the cocktail saved his life. In pictures from that trip, his eyes are hollowed out, his neck so thin that it juts from his woollen sweater like a broomstick. He got worse before he got better, he says—"It was like I had an anvil in my stomach." But one morning, about a month after his first dose, he woke up and felt like going for a walk. A few days after that, he had a strong urge to chop wood. He now takes three anti-retroviral and protease-inhibitor drugs every day and hasn't had a major medical problem in ten years. He still doesn't have the stamina he'd like, and his forehead is often beaded with sweat, even on cool evenings around the commune's dinner table. "I wish this weren't my reality," he told me. "I don't feel great that my life is medically managed. But if that's what's keeping me alive, hallelujah."

It's this part that incenses some of his readers: Having sung the praises of sauerkraut, revealed the secrets of kombucha, and gestured toward the green pastures of raw milk, Katz has surrendered to the false promise of Western medicine. His drug dependence is a sellout, they say—an act of bad faith. "Every two months or so, I get a letter from some well-meaning person who's decided that they have to tell me that I'm believing a lie," he told me. "That the H.I.V. is meaningless and doesn't make people sick. That if I follow this link and read the truth I will be freed from that lie and will stop having to take toxic pills and live happily ever after." Live cultures have been part of his healing, he said. They may even help prevent diseases like cancer. "But that doesn't mean that kombucha will cure your diabetes. It doesn't mean that sauerkraut cured my AIDS."

The trouble with being a diet guru, it seems, is that the more reasonable you try to be the more likely you are to offend your most fervent followers. *The Revolution Will Not Be Microwaved* includes a chapter called "Vegetarian Ethics and Humane Meat." It begins, "I love meat. The smell of it cooking can fill me with desire, and I find its juicy, rich flavor uniquely satisfying." Katz goes on to describe his dismay at commercial meat production, his respect for vegetarianism, and his halfhearted attempts to embrace it. "When I tried being vegan, I found myself dreaming about eggs," he writes. "I could find no virtue in denying my desires. I now understand that many nutrients are soluble only in fats, and animal fats can be vehicles of rich nourishment."

Needless to say, this argument didn't fly with much of his audience. Last year, the Canadian vegan punk band Propagandhi released a song called "Human(e) Meat (The Flensing of Sandor Katz)." Flensing is an archaic locution of the sort beloved by metal bands: it means to strip the blubber from a whale. "I swear I did my best to insure that his final moments were swift and free from fear," the singer yelps. "But consideration should be made for the fact that Sandor Katz was my first kill." He goes on to describe searing every hair on Katz's body, boiling his head in a stockpot, and turning it into a spreadable headcheese. "It's a horrible song," Katz told me. "When it came out, I was not amused. I had a little fear that some lost vegan youth would try to find meaning by carrying out this fantasy. But it's grown on me."

THE MOON WAS IN SAGITTARIUS on the last night of April, the stars out in their legions. Katz and I had arrived in the Smoky Mountains to join the gathering of the Green Path. About sixty people were camped on a sparsely wooded slope half an hour west of Asheville. Tents, lean-tos, and sleeping bags were scattered among the trees, below an open shed where meals were served: dandelion greens, nettle pesto, kava brownies—the usual. In a clearing nearby, an oak branch had been stripped and erected as a maypole, and a firepit dug for the night's ceremony: the ancient festival of Beltane, or Walpurgisnacht.

We'd spent the day going on plant walks, taking wildcrafting lessons, and listening to a succession of seekers and sages—Turtle, 7Song, Learning Deer. Every few hours, a cry would go up, and the tribe would gather for an adult version of what kindergartners call Circle Time: everyone holding hands and exchanging expressions of self-conscious wonder. The women wore their hair long and loose or bobbed like pixies'; their noses were pierced and their bodies wrapped in rag scarves and patterned skirts. The men, in dreadlocks and piratical buns, talked of Babylon and polyamory. The children ran heedlessly through the woods, needing no instruction in the art of absolute freedom. "Is your son homeschooled?" I asked one mother, who crisscrossed the country with her two children and a teepee and was known as the Queen of Roadkill. She laughed. "He's unschooled," she said. "He just learns as he goes."

The Green Path was part ecological retreat and part pagan revival meeting, but mostly it was a memorial for its founder, Frank Cook, who had died a year earlier. Cook was a botanist and teacher who travelled around the world collecting herbal lore, then writing and lecturing about it back in the U.S. He lived by barter and donation, refusing to be tied down by full-time work or a single residence, and was, by all accounts, an uncommonly gifted teacher. (He and Katz often taught seminars together.) As a patron saint, though, Cook had left his flock with an uneasy legacy. When he died, at forty-six, it was owing to a tapeworm infection acquired on his travels. Antibiotics might have cured him, but he mostly avoided them. By the time his mother and friends forced him to go to a hospital, last spring, his brain was riddled with tapeworm larvae and the cysts that formed around them. "Frank was pretty dogmatic about Western medicine," Katz said. "And I really think that's why he's dead."

Around the bonfire that night, I could see Katz on the other side of the circle, holding hands with his neighbors. After eighteen years in the wilderness, he couldn't imagine moving back to New York City—a weekend there could still wear him out. Yet his mind had never entirely left the Upper West Side, and his voice, clipped and skeptical, was a welcome astringent here. After a while, a woman stepped into the firelight, dressed in a long white gown with a crown of vines and spring flowers in her hair. Beltane was a time of ancient ferment, she said, when the powers of the sky come down and the powers of the earth rise up to meet them. She took two goblets and carried them to opposite points in the circle. They were full of May wine steeped with sweet woodruff, once considered an aphrodisiac. On this night, by our own acts of love and procreation, we would remind the fields and crops to grow.

There was more along those lines, though I confess that I didn't hear it. I was watching the kid to my left—a scruffy techno-peasant dressed in what looked like sackcloth and bark—take a swig of the wine. The goblets were moving clockwise around the circle, I'd noticed. By the time the other goblet reached me, thirty or forty people would have drunk from it.

I heard a throat being cleared somewhere in the crowd, and a cough quickly stifled. Embracing live cultures shouldn't mean sacrificing basic hygiene, Katz had told me that afternoon, after his sauerkraut seminar. "Part of respecting bacteria is recognizing where they can cause us problems." And so, when the kid had drunk his fill, I tapped his shoulder and asked for the goblet out of turn. I took a quick sip, sweet and bitter in equal measure. Then I watched as the wine made its way around the circle—teeming, as all things must, with an abundance of invisible life.

KRISTIN OHLSON

Earth on Fire

FROM *DISCOVER*

Around the world, thousands of underground seams of coal are on fire. Long ignored by scientists, these coal fires release toxins into the atmosphere, pollute groundwater, and add to global warming. Kristin Ohlson follows researchers who are focused on finding a way to combat this phenomenon.

NOT FAR FROM HAZARD, KENTUCKY, IN THE shadow of Lost Mountain, a woman named Ruth Mullins saw smoke rising off the slope. "I knew it wasn't no woods on fire, because of the smell"—the rotten-egg stench of sulfur—she said. Her suspicions were soon confirmed: Lost Mountain's coal mine, abandoned for 40 years, was burning.

Kentucky names coal fires for the people who first report them, so the fire, which has continued to smolder and occasionally flame since it was identified in 2007, is known officially as the Ruth Mullins fire. "We've never met the woman and we don't know where she lives, but her name now appears in scientific publications that are read all over the world," says Jennifer O'Keefe, a geologist at Kentucky's Morehead State University. "She's got her little bit of immortality."

O'Keefe is part of a team that has been visiting the Ruth Mullins fire over the past three years, studying its behavior and quantifying the gases that plume from nine known openings in the ground. Last January she and a colleague, University of Kentucky geologist James Hower, brought some students to the coal fire for new measurements. They parked off Highway 80, a road that cuts a swath along the side of Lost Mountain, and unloaded gear in a stingingly cold wind as speeding trucks whipped ice along the asphalt. Trudging up the snow-covered mountain, the scientists shivered along the flat shelf of land circling its midsection, the remains of contour mining in the 1950s. While smoke from the burning mine had been hard to spot from the road, here it billowed from small vents where portals to the mine had collapsed.

Approaching the site, all except Hower (who stayed farther back) donned pink respirators. A student equipped with a GPS device tried to detect the outline of the underground fire by looking for areas where the snow was thinner or melted away entirely. Two other students and O'Keefe settled at a vent, measuring the temperature at the opening and the velocity of the gases (including carbon monoxide, carbon dioxide, hydrogen sulfide, methane, and oxygen) that were flowing out.

"Jen, do we have any tar or minerals up there?" Hower called to O'Keefe, who shook her head. He made his way carefully over some fallen trees, possibly killed by the coal fire cooking their roots, to another vent and climbed closer, sliding a little in the snowmelt and mud made warm by the mine's hot breath. Here there was plenty of tar and minerals: Black goo two-toned the leaves on the ground, and minerals that had precipitated out of the gases encrusted the tree roots dangling over the vents. To identify the potentially dozens of hydrocarbon gases roiling beneath, he stuck a tube deep inside each vent, collecting emissions in a steel canister for later analysis in a laboratory at the University of California at Irvine.

Hower also retrieved a weathered contraption perched at the entrance to one of the vents. Cobbled together from galvanized-steel stovepipes and heat-resistant tape, this assemblage, nicknamed the Tin Man, had been taking measurements for 22 days. Three layers of filters impregnated with activated carbon captured mercury emissions, and a pair of instruments recorded temperature and carbon monoxide every 10 seconds for three days. Another set of devices monitored the same parameters every minute for the entire duration. Through these measurements, the team will gain a better understanding of the long-term variation in the fire's temperature and emissions.

This was the second Tin Man. The first, deployed during a 2009 study, showed that the carbon monoxide level at Ruth Mullins dropped dramatically once a day and then shot back up again. "These mine fires seem to have a regular breathing cycle," Hower says.

COAL FIRES ARE AS ANCIENT and as widely distributed as coal itself. People have reported fires in coal beds close to the earth's surface for thousands of years—in fact, Australia's Burning Mountain, once thought to be a volcano, sits atop a coal seam that has been on fire for some six millennia. But ever since the Industrial Revolution, the number of coal fires has grown dramatically. There are now thousands of such fires around the world, in every country—from France to South Africa to Borneo to China—where mining exposes coal deposits.

These fires are an insidious, persistent, and often nearly invisible threat to local health and to the natural and built environment. Added to that, there is now a growing realization that all these coal fires together may contribute significantly to climate change, a risk that has inspired the United States Geological Survey (USGS) to measure emissions of greenhouse gases and other pollutants from coal fires around the United States, starting with three in the Powder River Basin of Wyoming. The USGS effort, including scientists from organizations around the country, was convened to employ new tools and expertise to measure greenhouse gases from coal fires, which have not been included in previous national and worldwide surveys. "What is the overall contribution of these coal fires to global warming?" asks Glenn Stracher, a geologist at East Georgia College whose work inspired the USGS effort. "That's an important question that no one has answered, and that's why this team of scientists has gotten together to work on a quantitative analysis."

Most Americans are unaware of these long-burning coal fires, with the possible exception of the mine fire in Centralia, Pennsylvania. In 1962 residents of this small mining town burned trash in an abandoned strip mine used as a dump near the Odd Fellows Cemetery, not realizing that the mine had not been properly sealed. The trash was reduced to smoldering piles, which firefighters later extinguished—or so they thought. But the fire continued to burn, and a month later bulldozers arrived for a more concerted effort to put it out. The citizens then discovered that the dump contained a 15-foot-long opening that connected to a maze of underground mine tunnels. These passages allowed the fire to spread to the coal seam underneath the town and expand along four fronts, eventually affecting a surface area about two miles long and three-quarters of a mile wide.

Since then, around $4 million has been spent to put the Centralia fire out, to no avail. It continues to burn today, moving through a vast network of abandoned mines that are still littered and lined with coal. No one knows how extensive these empty spaces are, and the effort to quell the blaze has come to an end. "It's too expensive to tackle, and we're not sure we can do it anyway," says Alfred White-house, chief of the Reclamation Support Division of the federal Office of Surface Mining.

The town of Centralia is almost completely deserted today. After some residents passed out from carbon monoxide inhalation and another fell into the earth in 1981, when the ground suddenly collapsed—as the coal burns away, the ground above it often subsides into the resulting cavity—Pennsylvania received $42 million from Congress to relocate Centralia's residents. Folks accepted the buyout one by one, and their homes were demolished to discourage squatters. (Nine holdouts are still fighting eviction today.) The town now looks like a giant vacant parking lot. A few intersections still sport stop signs, which spray painters have modified to read "Don't STOP believing." Aside from the eerie emptiness, signs of the fire below are subtle. On a day in January, dead grasses bristle with ice along the edges of long cracks in the earth, and wisps of gas drift here and there. An area the size of a small house recently sank about three feet, and a bright green band of vegetation flourishes in the steaming, broken earth around it.

When Stracher first visited Centralia in 1991, the town looked even more like a disaster zone. Stracher had just finished his post-doctoral training in metamorphic petrology; as a new professor at Bloomsburg University of Pennsylvania, he went to Centralia on a geology field trip. He was horrified by the sinkholes encased in sulfur and other precipitated minerals, the huge cracks in now-abandoned Highway 61 near town, the thick fumes rising from a ravine called Death Valley, and the sulfur-laden trees around the ravine. The town's Catholic church was still standing then. Stracher posed for a photograph next to a mournful sign outside the church that read, "Centralia: Coal mine fire is our future."

"The fire had been burning so long by then," he recalls. "I wondered what long-term effect it was having on the atmosphere and groundwater, even on people who didn't live there."

At that point, Stracher did not know very much about coal, but he had a strong background in chemical thermodynamics. He decided to study the behavior of the sulfur coming out of Centralia's burning coal. In 1995 he reported that some of the sulfur crystallized and stayed on the ground, potentially tainting the local water, and some of it floated away as a gas, polluting the air. Nine years later, he and a former student published an article in the *International Journal of Coal Geology* titled "Coal Fires Burning Out of Control Around the World: Thermodynamic Recipe for Environmental Catastrophe." Over the next few years, Stracher was asked to put together symposia, one for the American Association for the Advancement of Science and another for the Geological Society of America. By that time, coal fires had become his life work.

After that first trip, Stracher quickly learned that Centralia was not America's oldest or biggest coal fire. It was not even Pennsylvania's oldest or biggest coal fire. At last count, the United States had 112 documented underground fires like Centralia and Ruth Mullins, along with many more yet to be counted. In addition to the underground fires, there are also 93 known surface coal fires, some of them in huge waste piles created during the process of coal mining. Stracher mentions a 100-foot-high burning "gob pile" (containing pieces of coal mixed with mudstone) near Birmingham, Alabama. The pile caught fire 20 years ago and was apparently extinguished at the time. But it reignited in 2006, emitting large amounts of smoke and toxic gases that caused respiratory complaints; the effort to extinguish it for good was just completed last March. Other surface fires occur where coal seams that sit close to the earth's crust are ignited by lightning strikes, forest fires, or brushfires. The coal-rich American West has a long history of such fires—in fact, the Powder River, whose basin in northeast Wyoming and southeast Montana is the source of about 40 percent of

America's coal, was so named because the area smelled like burning gunpowder.

But America's problem with coal fires is small compared with that of the rest of the world, where untold thousands of coal fires—no one can come up with a number judged to be even remotely accurate—burn unchecked. Eastern India has the densest concentration of coal fires in the world. Sixty-eight of them burn within a 174-square-mile region in the Jharia Coalfield in the state of Jharkand, some right next to areas where mining families live.

In China, estimates of the amount of coal consumed or made inaccessible by uncontrolled fires runs as high as 200 million metric tons per year, 10 percent of the country's total coal production. Indonesia, a major exporter of coal to the Pacific Rim, has many thousands of coal fires. Whitehouse spent several years there fighting the burn. In a 2004 paper, he stated that the number of coal fires in eastern Borneo might be as high as 3,000. Today he thinks even that estimate was far too low. "The real number is so astronomical that no one would believe it," he says. "The published numbers are about one percent of what could actually be there."

Most of the coal fires in Borneo start when local farmers and plantation owners burn brush to clear land for planting, accidentally igniting a coal seam just under the surface. Fires in both abandoned mines and waste piles sometimes start because of a nearby blaze, but they can also ignite through spontaneous combustion: Certain minerals in the coal, such as sulfides and pyrites, can oxidize and in the process generate enough heat to cause a fire.

GIVEN THE IMPLICATIONS FOR SAFETY, health, and climate, the paltry attention paid to coal fires puzzles and angers many. "Most of our efforts are unfunded or funded with a shoestring budget," says Anupma Prakash, a geologist at the University of Alaska with a career-long interest in coal fires. She, Stracher, and

Ellina Sokol of the Institute of Geology and Mineralogy, Siberian Branch of the Russian Academy of Sciences, are coeditors of Elsevier's four-volume *Coal and Peat Fires: A Global Perspective,* the first volume of which will be published this year. Even other scientists can find the issue obscure. "People ask me why they should worry about coal fires and want me to give them some numbers and hard facts, but the reliable quantitative data are not there at the moment," Prakash admits.

The USGS team wants to plug that gap, deploying ground sensors to tally the surface carbon dioxide emissions from a coal fire and then comparing the measurements with those from aerial surveys of the same fire using an infrared camera. By calculating the amount of burning coal needed to produce the hot spots picked up by the infrared, scientists can determine the amount of carbon dioxide such a fire should release. If both methods yield comparable measurements, the researchers will know they are closing in on solid data.

Last year the team spent three days clambering around the Powder River Basin, measuring gases from 29 vents at three fires. This alone would have given an incomplete assessment of emissions, because coal fires also release gases through the soil. So USGS geochemist Mark Engle built an "accumulation chamber" that measured gases coming out of the ground along a 119-point grid. He found that even in places where burning coal was so deep in the ground that there was no visual evidence on the surface, there were still significant amounts of carbon dioxide rising up. In fact, nearly as much CO_2 entered the atmosphere through the soil as from the vents. "The gas diffused out of the soil is not real obvious," Engle says. "The ground is not necessarily hot, and you can't trust the vegetation to tell you what's going on."

While the ground crew worked during the day, the airborne crew took off before sunrise; a cold, dark night provides the best contrast between the coal fires and the surrounding land. In the final analysis, the ground-based and airborne assessments of carbon dioxide agreed within one order of magnitude, close enough for the project to be considered a success. By showing that remote sensing not only can find coal fires but can accurately assess the amount of carbon dioxide and other gases being emitted, the USGS team has introduced an easier way of assigning solid numbers to the greenhouse gases emanating from coal fires around the world.

The team has yet to release its final numbers from the Wyoming study, but Hower and O'Keefe's earlier studies of the Ruth Mullins fire provide some sense of scale. Although the carbon contribution from each fire may seem modest, their prevalence and longevity add up. "They might not put out as much as a coal-burning power plant," O'Keefe says, "but they have usually been burning a lot longer and will keep burning a lot longer."

Coal fires also release a broader palette of noxious pollutants. When coal is burned in a power plant, operators make sure the fire gets plenty of oxygen so that it burns hot enough to produce the most possible energy and the fewest by-products. Coal burning in an abandoned mine typically gets far less oxygen. As a result, the coal smolders and releases a wide range of nasty, partially oxidized compounds. Testing at Centralia has revealed 45 organic and inorganic chemicals, including toxins like benzene, toluene, and xylene. Fifty-six compounds have been identified in the gases from one of China's coal fires.

"As the gases come to the surface, they react with rocks along the way and the chemistry is constantly changing," Stracher says. "It's very complex. We'll get a sample analyzed, and the chemist will say there are 40 to 60 compounds in it, and we have no idea what chemical reactions produced these compounds."

Depending on variables of chemistry, population, and ecology, health effects may be profound. According to O'Keefe, the Ruth Mullins fire presents a respiratory health hazard to the Route 80 area of Kentucky, while the state's Laura Campbell fire threatens the water supply. The gob fire in Alabama was the cause of traffic accidents. The Jharia Coalfield fire in India is responsible for cases of asthma, chronic bronchitis, and lung and skin cancer. One study suggests that coal fires in the United States may spit out as much as 11.5 metric tons of mercury annually, nearly a quarter as much as all the nation's coal-fired power plants. And unlike power plant emissions, coal fire emissions cannot be regulated or controlled.

MANY OF THE WORST COAL fires are in remote areas or in poor regions where ruined communities and impaired public health have not commanded much attention. Even in Centralia, the groundwater potentially tainted by the town's long-burning coal fire has never been tested. But with a growing recognition that coal-fire emissions may threaten the planet as a whole, scientists and others hope more resources will be deployed to put the insidious fires out—a task that is much harder than it sounds.

Extinguishing a fire requires cooling the coal and isolating it from both the heat and the oxygen that feeds combustion. Surface fires are the easiest to put out, with firefighters creating moats or breaks to keep the fire from spreading and then smothering it in a nonflammable material, most often clay. Slightly deeper fires can sometimes be quenched by digging out the burning coal—in Indonesia, Whitehouse's crews did this by hand—and then burying the entire area. But fires that rage deep underground, fed by oxygen coming from cracks in the earth, are extremely difficult to deal with. Most solutions involve pumping some combination of mud or fly ash combined with an inert gas or water, but the mixture does not always flow thoroughly enough to cover the burning coal, and it can crack when dried, allowing oxygen to get back in.

Stracher believes that one of the most promising approaches has been developed by a veteran Texas firefighter named Mark Cummins. Back in the 1980s Cummins developed an improved system to produce fire-fighting foam, and he has a long history of working with the material. In 2003 he used his nitrogen foam (made by condensing a laundry-detergent-like material) to act as a blanket, separating fire from the oxygen that fed it to help extinguish a blaze in West Virginia's Pinnacle Mine. Since then, Cummins has developed a foam that is loaded with microbes. After he puts out a fire with his original nitrogen foam, his plan is to shoot this second foam into the mine, where the microbes will consume oxygen and replace it with carbon dioxide. "At that point, you couldn't burn that mine if you set off a bomb down there," he says.

Up to now, interest in such expensive, little-tried approaches has been low. But Hower hopes that an expanded understanding of the environmental impact of these fires will change things. He notes that the Ruth Mullins fire is migrating slowly toward nearby Highway 80. If a coal seam burns through the road, asphalt could crack open and sink, swallowing people and cars and unleashing a hellish scenario that might finally make people pay attention to what is going on beneath their feet.

MICHAEL SPECTER

A Deadly Misdiagnosis

FROM *THE NEW YORKER*

> *While hardly a threat in the West, tuberculosis is still a killer in the developing world—and on the rise. Michael Specter visits India, which has some of the highest rates of TB in the world, to see whether new diagnostic tools and treatments that could halt the spread of the epidemic have a chance in such an impoverished country.*

EVERY AFTERNOON AT ABOUT FOUR, A SLIGHT WOMAN named Runi slips out of the cramped, airless room that she shares with her husband and their sixteen children. She skirts the drainage ditch in front of the building, then walks toward the pile of hardened dung cakes that people in this slum on the edge of the northeastern Indian city of Patna use for fuel. Dressed in a bright-yellow sari shot with gold threads, Runi is followed by several of her children. Although she can't remember their ages, or her own, Runi must be about forty, because she dates her life from its first crucial memory: the smallpox epidemic that devastated Patna and much of surrounding Bihar province in 1974.

Runi survived that plague, and several others, but, about a year ago, after developing a persistent cough, she visited one of the private medical clinics that line the streets of Patna. There someone who called himself a doctor stuck a needle in her arm, drew a few drops of blood, examined them, and told her that she had tuberculosis. It is not an uncommon diagnosis. Tuberculosis has always been the signature disease of urban poverty, passed easily in poorly ventilated spaces. India has nearly two million new cases each year, and every day a thousand people die of the disease, the highest number in the world. Tuberculosis is also the leading cause of death among people between fifteen and forty-five—the most productive age group in any country and the key to India's prospects for continued economic growth.

For most patients, the choices are bleak. Public hospitals are so overcrowded that people are forced to rely on inaccurate tests dispensed at private labs and clinics. They are unregulated enterprises, and peddle blood tests that are responsible for tens of thousands of misdiagnoses every year. "This is deadly," L. S. Chauhan, the director of the National TB Control Program, told me when we met in New Delhi. "But there are thousands of labs. Shut one down and the next day ten more appear."

Runi's test was indeed worthless. It determined the presence of antibodies, which show that a body's immune system has begun to respond to an infection. But most TB infections are latent: no more than ten per cent will ever cause illness. This means that ninety per cent of people with antibodies for TB in their blood don't have the disease. Runi's cough was clearly caused by something else.

Vaccines and antibiotics have long been seen as touchstones of medical progress. To stop tuberculosis, however, particularly in the developing world, an accurate diagnostic exam is needed even more. In India, China, and Africa, at least two billion people have latent infections. Yet every day thousands are told, mistakenly, that they are sick and need treatment. That's what happened to Runi. Soon after she received her diagnosis, Runi began a regimen of powerful (and toxic) drugs provided by the public-health service, and she stuck to the program for the required six months. Not long after finishing, however, she started to feel worse than she ever had before. "This is the tragedy of our TB-control program," Shamim Mannan said as we watched Runi's children play. Mannan, who is from Assam, a few hundred miles from Patna, serves as the Indian government's chief TB consultant in the region.

"Officially, she is cured," he said. "But how would we know? She took a test that showed she had the antibody for TB in her blood. So do I. So do five hundred million Indians." As Runi stooped to gather fuel for the stove, she began to cough, lightly at first and then with alarming force. Every cough sounded as if somebody had shattered a pane of glass.

"Now she really is sick," he continued, explaining that Runi's TB was no longer dormant, and that taking drugs when they are not necessary often makes them ineffective when they are. "This is what happens when tests mislead us. She will need the drugs again. If they don't work properly, she will be in real trouble. She has almost certainly infected some of her children. That makes everything harder, more expensive, more painful."

TUBERCULOSIS STRIKES VULNERABLE PEOPLE WITH special ferocity. Victims are seized by severe night sweats, wasted by fatigue, and punished by the blood-tinged cough that is the disease's defining symbol. In most cases, tuberculosis affects the lungs, but it can invade almost any organ of the body. When an infectious person coughs, sneezes, spits, or even shouts, he sends minute particles of sputum, or phlegm, into the air—exposing anyone nearby. For many years, the disease, which is caused by *Mycobacterium tuberculosis*, was referred to as "consumption," because without effective treatment patients often wasted away.

To fight the infection, the body's immune system forms a scar around the TB bacteria which serves as a kind of moat. Afterward, the bacteria lie dormant and cannot spread or infect others. But immune systems fail, and when that happens TB can move from the lungs to the bloodstream and then to the kidneys, the brain, and other organs. (That's why in patients with H.I.V., which ravages the cells that the body uses to defend itself, tuberculosis becomes particularly deadly.) The only way to cure the disease is with a combination of antibiotics. The treatment lasts six months because the drugs work only when the TB bacteria—which grow slowly—are dividing.

For centuries, tuberculosis has been the source of misguided stereotypes, including the association of consumption with creativity and brilliance. "Doctors suspect that tuberculosis develops genius," a 1940 article in *Time* pointed out, "because 1) apprehension of death inspires a burning awareness of life's beauty, significance, transience, 2) the bacillus breeds restlessness and an intoxicated hypersensitiveness." Keats, Chekhov, the Brontë sisters, and George Orwell—who was born not far from Patna, where his father managed the regional opium trade—all died of the disease.

Nonetheless, tuberculosis has always taken its most serious toll on the industrial-labor class—not on artists. The rise of industry throughout the world has been mirrored uncannily by a rise in deaths from tuberculosis. It was the leading cause of death in Europe and the United States from the eighteenth century into the twentieth. Then prosperity—rather than medicine—drove the rate of infection down. As a society becomes richer, the conditions that allow tuberculosis to flourish start to wane. Sanitation and housing improve and so does nutrition. By the nineteen-fifties, very few people in the West were dying of the disease.

In the developing world, though, tuberculosis has surged dangerously, and this year, according to the World Health Organization, there will be ten million new cases, the largest number in history. As people join the great migrations from villages to crowded cities, slum life and tuberculosis await them. With India's urban population expected to double in the next thirty years, to seven hundred million, its cities will remain fertile ground for an infectious epidemic. Yet—no doubt owing to the fact that rich people in the West rarely get the disease—tuberculosis receives fewer resources, fewer research dollars, and less attention from the global health community than either AIDS or malaria—the two other most deadly infectious diseases. TB activists don't march on Washington or chain themselves to the gates of pharmaceutical firms to demand better treatment.

Tuberculosis can be cured, but taking several antibiotics nearly every day for six months is not easy, particularly in parts of the world without running water or refrigeration. In 1994, the W.H.O. instituted a program called DOTS, which stands for "directly observed treatment, short course." DOTS requires health workers to provide medicine—and then to watch people swallow it every day until they complete their treatment. Compliance is essential, because stopping treatment in the middle permits the most resilient strains of the bacteria to thrive, greatly increasing the chance that they will become resistant to basic, inexpensive drugs.

Thirty-six million people have received care under the DOTS program, eight million of whom would have died without it. It has been a triumph by any measure. Even DOTS, though, has not been able to keep the disease from spreading. That is largely because there is no cheap, reliable test that can determine who is sick and who is not.

Blood tests, like the one Runi had, often do more harm than

good. One recent study found that Indians undergo more than 1.5 million useless TB tests of this kind every year. Other approaches are almost as unreliable. Examining a person's sputum—a diagnostic procedure that was developed more than a century ago—remains the most common way to detect the infection. It is a laborious process. Technicians smear the sputum on a slide and then place the specimen under a microscope. The instructions are comically complex. "Spread sputum on the slide using a broomstick," a typical recipe, posted on the wall of a clinic in Patna, begins. "Allow the slide to air dry for fifteen to thirty minutes. Fix the slide by passing it over a flame from three to five times for three to four seconds each time." If the slide isn't held over the flame long enough, false stains will appear—suggesting that people are sick when they are not. Hold the slide too long, though, and the stain will disappear and show nothing at all. The results are accurate little more than half the time.

"You can treat a lot of people, and India has," said Madhukar Pai, an epidemiologist at McGill University and the co-chairman of the international group that assesses new diagnostics for the Stop TB Partnership. "But if you have tests that cause misdiagnosis on a massive scale you are going to have a serious problem. And they do."

MEDICINE RARELY PROVIDES MAGIC BULLETS, but, for the first time, a technology has been developed that might help countries like India escape the endless cycle of mistaken diagnoses and haphazard treatment. A company called Cepheid, based in Sunnyvale, California, now makes a device, called a GeneXpert, that allows doctors to diagnose TB in under two hours—without error or doubt. "The machine is so powerful that it could help end tuberculosis," Mannan told me. "I don't think that is an exaggeration."

An editorial three months ago in the *New England Journal of Medicine* also raised the possibility that, with proper use of this device, tuberculosis—a disease that has been around since the days of the Pharaohs—could be eliminated. The cost, however, would be far too high for the Indian Ministry of Health. "Private business would have to take the lead," Mannan said. "In the past, countries waited until they got richer and tuberculosis mostly went away. India cannot do that. The epidemic is just too big. And we are too poor."

The GeneXpert was developed in 2002, with initial support from the Department of Defense. After the events of September 11th and the mailing of anthrax spores later that year, biological threats became a national priority. The only sure way to recognize dangerous new organisms, whether made by man or by nature, is to analyze their unique DNA, and the GeneXpert has tested billions of pieces of mail for toxins. Its diagnostic capabilities seemed even more promising, however. In 2008, with funding from the Bill and Melinda Gates Foundation, the Foundation for Innovative New Diagnostics, and the National Institutes of Health, researchers at medical centers throughout the world began to assess the machine's effectiveness in diagnosing tuberculosis.

Its success was striking. In a study published along with that editorial in the *Journal*, researchers reported that the GeneXpert identified more than ninety-eight per cent of active TB infections, including many that sputum smears had missed. Because the test looks for the TB bacterium itself, rather than for antibodies, latent infections don't confuse the GeneXpert as they do blood tests. The machine costs nearly twenty-five thousand dollars and each test is about twenty dollars. Prices could plunge if similar machines were introduced and used widely.

"This is absolutely transformational technology," Peter Small, the director of tuberculosis programs for the Gates Foundation, said. "It is a system that removes the guesswork from one of our most deadly diseases." Unlike the sputum technique, the molecular approach is straightforward: a patient spits into a cup, and the sample is placed in a cartridge that looks much like the pods used in many espresso machines. A computer examines the sample's DNA to see if it contains the genetic signature of TB. Results are available within hours.

The GeneXpert can even determine whether the bacteria are resistant to rifampicin, the most effective and widely used component of the four-drug cocktail commonly prescribed for TB. "People often equate sophisticated science with complexity, and this is just the opposite," Small said. "As long as there is electricity, the tests could be carried out by unskilled workers in any village. Training them would be easy, and the potential benefits—saving billions of dollars and millions of lives—worth any effort. The question is how do we get there. I have heard people say that we should trust the government bureaucracy. But others say let's put our faith in an unregulated collection of free agents. It's hard to know which approach is more ludicrous."

I put that question to Mannan, the official responsible for TB control in the Bihar region. A slight, intense man with eyes the color of wet coal, Mannan is a former Army doctor who left the service after he injured his leg jumping from an airplane. He has been frustrated by how rarely the promise of Indian medicine is realized, and by how little entrepreneurs—in one of the world's most entrepreneurial countries—are doing to help.

"We do know that private enterprise can work in India," he said. "Just look at the mobile-phone industry. And the public efforts to halt major diseases have been remarkable. But how do we get them to work together?" Nobody has an answer to that question. The interplay between public and private medicine in India is difficult to navigate, in part because the quality of private medicine varies so wildly. To demonstrate the range of medical options open to most people in Bihar, Mannan suggested that we travel to Darbhanga, about ninety miles northeast of Patna. Before we left, he said, "Everything you find in the country, the good and the bad—it is all in Darbhanga."

* * *

EVEN AT FIRST LIGHT, THE road that leads from Patna, Bihar's capital, to Darbhanga is impossibly crowded. On the ramp of the Mahatma Gandhi Bridge, which passes over the Ganges and leads north toward Nepal, oxen jostle with motorcycles and giant trucks. On the day I made the trip, the traffic was so heavy on the bridge—at more than three and a half miles, it's one of the longest in the world—that it took an hour just to reach the lush banana plantations on the other side.

Patna and Darbhanga were once important centers of civilization. Buddha found enlightenment under a bodhi tree in Bihar, twenty-five hundred years ago, and the Fortress of Maharajas still stands in Darbhanga. Today, though, the province lags behind other regions of India in every category of economic and human development. Its eighty-five million residents earn, on average, less than half what people in the rest of the country earn; plumbing and sanitation facilities are meagre. Tens of thousands of migrants pass through Darbhanga each year as they abandon their ancestral villages and seek new lives in Delhi, Mumbai, and other major cities.

THE MEDICAL COLLEGE HOSPITAL, AN imposing white fortress spread over several city blocks, is the largest in the region, but the city is also home to what may well be India's most formidable collection of unregulated pharmaceutical wholesalers, a kind of medical red-light district. Virtually any drug can be purchased, in whatever quantity one desires, without a prescription. Want a thousand polio vaccines? Narcotic painkillers, cancer medication? Scarce AIDS therapies? They are all readily available in Darbhanga. But rarely at the hospital.

The tuberculosis and AIDS clinics at the Medical College Hospital are open every day from 8 A.M. to 2 P.M. By the time Mannan and I walked into the cavernous waiting room early that morning, pa-

tients packed the benches and sprawled across the floor. Most sat silently, their eyes hollow, their heads down. The sound of harsh
coughing filled the air. The line for medications snaked into the
courtyard, where dozens of women, many of them cradling infants
in their arms, waited patiently.

Like other public hospitals in the developing world, the Medical
College Hospital struggles to provide medicine for its patients. The
dispensary is rudimentary: basic tuberculosis drugs are available,
but not those needed to treat resistant strains, which now account
for nearly twenty per cent of India's growing caseload. For people
who do not respond to the first line of TB treatments, there are two
choices: find money to buy medicine somewhere else or get sicker.

Since late 2009, the hospital has had one unique asset: a piece of
equipment called a P.C.R., which can multiply tiny samples of DNA
and analyze them. The device is not as fast as the GeneXpert, but it
can examine the genetics of virtually any organism, including tuberculosis. The hospital's machine, which was purchased with money
from a government research grant, has never been used. "The hospital has had this for months," Mannan said. "But nobody knows how
it works." We were standing at the door of the virology lab, where the
new P.C.R. Cobas TaqMan 48, made by Roche and sold for roughly
fifty thousand dollars, was resting on a shelf, still wrapped in its
shipping material.

How could that be? I was staring at a machine that could alter,
even save, the lives of scores of the people who were sitting nearby in
the gathering heat. Mannan said nothing, though his anger was pal-
pable. "Ask them," he said, referring to the scientists who worked in
the hospital, when I tried to get him to explain. "They will tell you."

We walked down the hall to meet Ravindra Prasad, a doctor in
the department of social medicine. He was an agreeable man with a
round face and an easy manner. I asked why the P.C.R. machine sat
imprisoned and unused.

* * *

"THE CHEMICAL KIT EXPIRED," HE said, smiling politely. The chemicals used in the machine have a short shelf life; but I learned later that they are not hard to replace. That couldn't have been the reason. "The methods we have for diagnosing tuberculosis all function smoothly," Prasad added, as if he were reading from a prepared statement. He was referring to sputum tests, which are often inaccurate. "We follow the standard manual." Prasad offered us tea, but said nothing more about the medical needs of his patients. "It's a nice lab," Mannan said when we left. "Beautiful, actually. But if the doctors used it properly that would interfere with their private practice."

I asked what he meant.

"It is simple," he said. "If patients are treated at the hospital, they won't need to pay for anything else."

THE DARBHANGA MEDICAL RED-LIGHT DISTRICT lies just a few blocks from the main hospital. On most days, as the public clinics prepare to take their last patients, touts appear in the waiting rooms and on the hospital grounds, eager to steer people toward a private doctor on Hospital Road. More than eighty per cent of medical services in India are in private hands, and health-care costs are among the most common reasons for bankruptcy.

The touts—equal parts salesmen, psychologists, and pimps—are good at their job. If you need TB medication or a test or an X-ray, these men will get you quickly to a clinic that charges for services people are entitled to receive at no cost in public hospitals. According to Mannan, the tout receives ten per cent of any eventual fee from a referral. Rickshaw drivers get five per cent, medical assistants ten, and the referring doctor, almost always a physician based at the Medical College Hospital, thirty-five per cent. That leaves forty per cent for the clinician.

Much of the time, the referring physician from the public hospital is also the private clinician who does the work. That earns him seventy-five per cent of any fee. Public salaries are not sufficient to support most doctors, so, every afternoon, many of the hospital's physicians work in these private clinics.

Well-trained doctors are not the only people working on Hospital Road, however. Officially, a doctor needs a license to practice medicine in India. In fact, though, there are no mechanisms to verify the validity of licenses or to punish people who break the law. It is not rare for "doctors" to lack medical training completely.

We arrived as darkness began to fall; hundreds of people, having finished the workday, crowded the rutted streets. There were dozens of drug shops, with names like Raj Medical Agency, Krishna Scientific and Surgical Works, and Zar Whole Sale Drugs—often illuminated by a single bulb. The streets of the medical red-light district are filled with "specialists." Mannan and I wandered into a back alley where two men asked after our health with more solicitousness than was necessary. I asked what they were offering, and one of them let out a loud cackle.

"Let me show you," he said, and led us to a small room with several chairs, a table, and three refrigerators. The man said that his name was Pranay, and he offered a variety of blood tests, for liver function, kidney function, H.I.V., and several other standard diagnostics, all at reasonable prices. Wholesalers make their money through volume sales, not high prices. "We get twenty-five to thirty referrals a day," he told me.

The stall next door could have been an exhibit in a science museum: it contained an ancient X-ray machine, held together with duct tape and baling wire. The owner had just finished taking chest slides for a middle-aged man. He didn't offer any of the customary lead shields or other protections against possible radiation leaks—and that machine certainly leaked. "It's safe," the man said. "They are X-rays."

He told us that he ran about fifteen to twenty chest X-rays a day; he charges a hundred rupees for each, or a little more than two dollars. His services were also available for broken bones and other routine problems. I asked how he had acquired his equipment and where he had learned to use it. He told us that he had taken the X-ray machine from a hospital in Bihar that was about to throw it away. The idea of training made him laugh. "Did you see 'Slumdog Millionaire'?" he said. "Before this, I was a chai wallah"—a man who serves tea—"just like that kid."

It was time to return to Patna; driving late at night on the roads of rural India is a risky business. Before we left, though, Mannan insisted that we make one more stop, at another clinic nearby. The place was essentially an open concrete garage; against one wall stood a small table with hot plates on which patients could heat rice. The room was full, and more than a dozen people stood on the street, waiting to get in. "This is the best TB clinic in town," a pharmacist who owned the shop next door explained.

The head of the clinic, Dr. P. M. Srivastav, works at Medical College Hospital, and we had spoken with him earlier. At night, for a hundred and thirty rupees, Srivastav will see anyone who waits in line. He doesn't test for tuberculosis at his clinic, and said that he refers people he suspects of having the disease to the hospital. He does, however, earn a fee from every patient he sees, including those he sends back to the hospital for free treatment. "Now do you understand why that machine is wrapped in plastic?" Mannan asked.

As we were about to leave, a large car pulled up at the front door. Srivastav climbed out of the back seat, looked at us with surprise, and smiled sheepishly. Before I had a chance to ask a question, he was gone, safely tucked away in his private office.

THE UNCERTAINTIES AND DANGERS OF diagnosis remain the greatest obstacle to successful TB treatment, in India and throughout the developing world. For that to change, investments from international aid organizations and from private companies will be necessary. That may seem unlikely, but it has happened before, most notably with AIDS drugs. In the nineteen-eighties, when AZT became the first effective treatment for H.I.V., the annual cost for each patient was ten thousand dollars. People in the West, who were rich or lucky enough to have good insurance, could afford it. In countries that struggle to provide basic immunizations against diseases like measles, though, AIDS treatments were a fantasy. Then various groups, including the Clinton Foundation, the Gates Foundation, and the Global Fund to Fight AIDS, Tuberculosis and Malaria, joined together to push for lower prices. Generic manufacturers, led by Cipla, the Mumbai-based pharmaceutical giant, began to churn out highly effective medicine at a small fraction of what it cost in the United States. Political pressure mounted, officials of the World Health Organization joined the call for cheaper AIDS medications, and today the governments of poor countries like India can buy those drugs for an annual price of less than a hundred dollars per patient. These drugs are normally distributed in bulk, through international AIDS organizations.

A similar effort will be required to lower the cost of diagnosing tuberculosis. There will also have to be a transformation in how TB medicine is regulated. That may seem like an insurmountable barrier, but, with the proper incentives, the system could work. Again, one can look to the history with AIDS medicines for a model. Because Cipla and other Indian pharmaceutical companies are frequently inspected by international regulators—such as the U.S. Food and Drug Administration—governments are willing to buy their products. That's one reason that Indian firms have become the most important manufacturers of generic AIDS medicines in the world.

Any company that sells molecular diagnostics would need the same sort of oversight. But producing cheap, internationally acceptable versions of the GeneXpert would surely lead to great profits.

"You have to keep in mind that India has many terrible doctors," Madhukar Pai told me. "But it also has some of the best private medicine available." I saw that in Darbhanga, where, in addition to the shoddy purveyors of the medical red-light district, I visited the Geeta Molecular Diagnostic Lab, a new private facility not far from the center of town. There I was greeted by a team of researchers, all in starched lab coats, including Deepak K. Prasad, a geneticist and the director of the laboratory. He led us on a tour: there were separate sections for gene detection, gene amplification, and histological analysis. Geeta Diagnostics had two P.C.R. machines and other, similarly advanced diagnostic tools. Few facilities in New York are better equipped. Patients sat on cream-colored couches reading magazines and sipping tea.

"The genetic approach to diagnosis is really where medicine is going," Prasad told us. The company, which is two years old, offers tests for heart disease, several types of cancer, thyroid disease, H.I.V., and tuberculosis, among other disorders. The TB test costs fifteen hundred rupees—a little more than thirty dollars. The lab does between fifty and seventy-five each week, and its doctors are paid well enough so that they don't need to work at second jobs.

"You can call it expensive, but you have to look at the eventual costs, not the initial price of a single test or one piece of machinery," Prasad said. That would be difficult to dispute. Thirty dollars may be a lot of money for most Indians; but treating drug-resistant strains of tuberculosis costs thousands of dollars and places a terrible burden on the country, not to mention on the people who are sick. In fact, treatment and deaths caused by TB in India cost more than three billion dollars in lost productivity each year.

The power of machines like the GeneXpert has already become evident at Mumbai's Hinduja Hospital, a private institution that has been using one for three years. Mumbai has one of the worst TB problems in India, particularly with drug-resistant cases. Yet at Hinduja the machine has made it possible for doctors to diagnose and treat patients before they are able to spread the disease. "There has always been a pretty standard approach to using fancy medical technology," Camilla Rodrigues, who runs the microbiology department, told me when I visited. "You develop it in the West and use it there. Eventually, it trickles down to the poor countries." Rodrigues pointed out that, with tuberculosis, the pattern makes no sense. The GeneXpert was invented in the West, but India and Africa need it much more urgently. "Every time we make a correct diagnosis, we save not one life but many," she said, waving in the direction of the boxy metal-and-Plexiglas machine sitting in a corner of the lab. "And with this machine we make correct diagnoses in two hours."

Rodrigues has been working with tuberculosis for two decades. "When I started, it seemed hopeless," she said as we sat in her office, which is adjacent to a busy lab filled with graduate students, most of whom are focussing on TB.

"You would ask people why we are not doing more to stop this terrible, crippling epidemic, and the answer was usually a shrug," she continued. Rodrigues has cavernous eyes and long dark hair pulled back in a bun. She speaks frankly but somehow conveys a buoyant sense of optimism. "For so long tuberculosis has been a part of life here. In the past, if you said you have the disease people would hardly flinch. Can you imagine going to a neighbor in New York and saying you have tuberculosis? People would shriek."

Lately, though, Rodrigues has begun to sense a shift away from the habitual fatalism that has defined the Indian approach to public health. "Sometimes I go to Churchgate Station," she continued. "It is the busiest train station in the city, maybe in the country. I go at rush hour. You cannot move or breathe or think. You cannot walk or talk. It is the perfect place to spread tuberculosis.

"But it is also the perfect place to stop it," she said. "I walk around that platform and I look at people and I say to myself, Which of you are sick? We need to know. And, finally, after more than a century we can know. At this point, it is just a matter of will."

MARK BOWDEN

The Enemy Within

FROM *THE ATLANTIC MONTHLY*

> *The Conficker worm, which has infected millions of computers worldwide, remains an enigma—and a threat. Mark Bowden reports on the Internet detectives who are trying to solve the mystery of Conficker's origins and ultimate intentions.*

T HE FIRST SURPRISING THING ABOUT THE WORM
that landed in Philip Porras's digital petri dish 18 months ago
was how fast it grew.

He first spotted it on Thursday, November 20, 2008. Computer-security experts around the world who didn't take notice of it that first day soon did. Porras is part of a loose community of high-level geeks who guard computer systems and monitor the health of the Internet by maintaining "honeypots," unprotected computers irresistible to "malware," or malicious software. A honeypot is either a real computer or a virtual one within a larger computer designed to snare malware. There are also "honeynets," which are networks of honeypots. A worm is a cunningly efficient little packet of data in computer code, designed to slip inside a computer and set up shop without attracting attention, and to do what this one was so good at: replicate itself.

Most of what honeypots snare is routine, the viral annoyances that have bedeviled computer-users everywhere for the past 15 years or so, illustrating the principle that any new tool, no matter how useful to humankind, will eventually be used for harm. Viruses are responsible for such things as the spamming of your inbox with penis-enlargement come-ons or million-dollar investment opportunities in Nigeria. Some malware is designed to damage or destroy your computer, so once you get the infection, you quickly know it. More-sophisticated computer viruses, like the most successful biological viruses, and like this new worm, are designed for stealth. Only the most technically capable and vigilant computer-operators would ever notice that one had checked in.

Porras, who operates a large honeynet for SRI International in Menlo Park, California, noted the initial infection, and then an immediate reinfection. Then another and another and another. The worm, once nestled inside a computer, began automatically scanning for new computers to invade, so it spread exponentially. It exploited a flaw in Microsoft Windows, particularly Windows 2000, Windows XP, and Windows Server 2003—some of the most common operating systems in the world—so it readily found new hosts. As the volume increased, the rate of repeat infections in Porras's honeynet accelerated. Within hours, duplicates of the worm were crowding in so rapidly that they began to push all the other malware, the ordinary daily fare, out of the way. If the typical inflow is like a stream from a faucet, this new strain seemed shot out of a fire hose. It came from computer addresses all over the world. Soon

Porras began to hear from others in his field who were seeing the same thing. Given the instant and omnidirectional nature of the Internet, no one could tell where the worm had originated. Overnight, it was everywhere. And on closer inspection, it became clear that voracity was just the first of its remarkable traits.

Various labs assigned names to the worm. It was dubbed "Downadup" and "Kido," but the name that stuck was "Conficker," which it was given after it tried to contact a fake security Web site, trafficconverter.biz. Microsoft security programmers shuffled the letters and came up with *Conficker*, which stuck partly because *ficker* is German slang for "motherfucker," and the worm was certainly that. At the same time that Conficker was spewing into honeypots, it was quietly slipping into personal computers worldwide—an estimated 500,000 in the first month.

Why? What was its purpose? What was it telling all those computers to do?

Imagine your computer to be a big spaceship, like the starship *Enterprise* on *Star Trek*. The ship is so complex and sophisticated that even an experienced commander like Captain James T. Kirk has only a general sense of how every facet of it works. From his wide swivel chair on the bridge, he can order it to fly, maneuver, and fight, but he cannot fully comprehend all its inner workings. The ship contains many complex, interrelated systems, each with its own function and history—systems for, say, guidance, maneuvers, power, air and water, communications, temperature control, weapons, defensive measures, etc. Each system has its own operator, performing routine maintenance, exchanging information, making fine adjustments, keeping it running or ready. When idling or cruising, the ship essentially runs itself without a word from Captain Kirk. It obeys when he issues a command, and then returns to its latent mode, busily doing its own thing until the next time it is needed.

Now imagine a clever invader, an enemy infiltrator, who *does* understand the inner workings of the ship. He knows it well enough to find a portal with a broken lock overlooked by the ship's otherwise vigilant defenses—like, say, a flaw in Microsoft's operating platform. So no one notices when he slips in. He trips no alarm, and then, to prevent another clever invader from exploiting the same weakness, he repairs the broken lock and seals the portal shut behind him. He *improves* the ship's defenses. Ensconced securely inside, he silently sets himself up as the ship's alternate commander. He enlists the various operating functions of the ship to do his bidding, careful to avoid tripping any alarms. Captain Kirk is still up on the bridge in his swivel chair with the magnificent instrument arrays, unaware that he now has a rival in the depths of his ship. The *Enterprise* continues to perform as it always has. Meanwhile, the invader begins surreptitiously communicating with his own distant commander, letting him know that he is in position and ready, waiting for instructions.

And now imagine a vast fleet, in which the *Enterprise* is only one ship among millions, all of them infiltrated in exactly the same way, each ship with its hidden pilot, ever alert to an outside command. In the real world, this infiltrated fleet is called a "botnet," a network of infected, "robot" computers. The first job of a worm like Conficker is to infect and link together as many computers as possible—the phenomenon witnessed by Porras and other security geeks in their honeypots. Thousands of botnets exist, most of them relatively small—a few thousand or a few tens of thousands of infected computers. More than a billion computers are in use around the world, and by some estimates, a fourth of them have been surreptitiously linked to a botnet. But few botnets approach the size and menace of the one created by Conficker, which has stealthily linked between 6 million and 7 million computers.

Once created, botnets are valuable tools for criminal enterprise. Among other things, they can be used to efficiently distribute malware, to steal private information from otherwise secure Web sites or computers, to assist in fraudulent schemes, or to launch denial-of-service attacks—overwhelming a target computer with a flood of requests for response. The creator of an effective botnet, one with a wide range and the staying power to defeat security measures, can use it himself for one of the above scams, or he can sell or lease it to people who specialize in exploiting botnets. (Botnets can be bought or leased in underground markets online.)

Beyond criminal enterprise, botnets are also potentially dangerous weapons. If the right order were given, and all these computers worked together in one concerted effort, a botnet with that much computing power could crack many codes, break into and plunder just about any protected database in the world, and potentially hobble or even destroy almost any computer network, including those that make up a country's vital modern infrastructure: systems that control banking, telephones, energy flow, air traffic, health-care information—even the Internet itself.

The key word there is *could*, because so far Conficker has done none of those things. It has been activated only once, to perform a relatively mundane spamming operation—enough to demonstrate that it is not benign. No one knows who created it. No one yet fully understands how it works. No one knows how to stop it or kill it. And no one even knows for sure *why* it exists.

If yours is one of the infected machines, you are like Captain Kirk, seemingly in full command of your ship, unaware that you have a hidden rival, or that you are part of this vast robot fleet. The worm inside your machine is not idle. It is stealthily running, issuing small maintenance commands, working to protect itself from being discovered and removed, biding its time, and periodically checking in with its command-and-control center. Conficker has taken over a large part of our digital world, and so far most people haven't even noticed.

The struggle against this remarkable worm is a sort of chess match unfolding in the esoteric world of computer security. It pits the cleverest attackers in the world, the bad guys, against the cleverest defenders in the world, the good guys (who have been dubbed the "Conficker Cabal"). It has prompted the first truly concerted global effort to kill a computer virus, extraordinary feats of international cooperation, and the deployment of state-of-the-art decryption techniques—moves and countermoves at the highest level of programming. The good guys have gone to unprecedented lengths, and have had successes beyond anything they would have thought possible when they started. But a year and a half into the battle, here's the bottom line:

The worm is winning.

A DIGITAL SAM SPADE

Twenty years ago, computers were bedeviled by hackers. These were savvy outlaws who used their deep knowledge of operating systems to invade, steal, and destroy, or sometimes just to tap into secure facilities and show off their skills. Hackers became heroes to a generation of teenagers, and had all sorts of motives, but their most distinctive trait was a tendency to show off.

Some had truly malicious intent. In his 1989 best seller, *The Cuckoo's Egg*, Cliff Stoll told the story of his stubborn, virtually single-handed hunt for an elusive hacker in Germany who was using Stoll's computer system at the Lawrence Berkeley National Laboratory as a portal to Defense Department computers. For many people, Stoll's book was the introduction to the netherworld of rarefied gamesmanship that defines computer security. Stoll's hacker never penetrated the most secret corners of the national-security net, and even relatively serious breaches like the one Stoll described were more nuisance than threat. But the individual hacker working as a spy or vandal has evolved into something more organized and menacing.

Andre' M. DiMino, a computer sleuth who is part of the Conficker Cabal, is considered one of the world's foremost authorities on botnets. He stumbled into his avocation on a Monday morning a decade ago, when he discovered that over the weekend, someone had broken into the computer system he was administering for a small company in New Jersey. DiMino has an undergraduate degree in electrical engineering with an emphasis in computer science, but he has mostly taught himself up to his present level of expertise, which is extreme. At 45, he is a slender, affable idealist who keeps a small array of computers in an upstairs bedroom. When I stopped by to talk to him, he baked me pizza. His day job is doing computer forensics for law enforcement in Bergen County, New Jersey, but he has a kind of alter ego as what he calls a "botnet hunter."

Back when he discovered the weekend break-in, DiMino assumed at first that it was the work of a hacker, a vandal, or possibly a former employee, only to discover, based on an analysis of the IP (Internet Protocol) addresses of the incoming data, that his little computer network had been invaded by someone from Turkey or Ukraine. What would someone halfway around the planet want with the computer system of a small business-management firm in a New Jersey office park? Apparently, judging by what he found, his invader was in the business of selling pirated software, movies, and music. Needing large amounts of digital storage space to hide stolen inventory, the culprit seemed to have conducted an automated search over the Internet, looking worldwide for vulnerable systems with large amounts of unused disc space—DiMino equates it to walking around rattling doorknobs, looking for one door left unlocked. DiMino's system fit the bill, so the crooks had dumped a huge bloc of data onto his discs. He erased the stash and locked the door that had allowed the pirates in. As far as the company was concerned, that solved the problem. No harm done. No need to call the police or investigate further.

But DiMino was intrigued. He reviewed the server logs for previous weeks and saw that this successful invasion was one of many such efforts. Other attackers had been rattling the doors of his network, looking for vulnerabilities. If there were bad guys actively exploiting other people's computers all over the world, designing sophisticated programs to exploit weaknesses . . . how cool was that? And who was trying to stop them?

DiMino set about educating himself on the fine points of this obscure battle of wits. He eventually co-founded the Shadowserver Foundation, a nonprofit partnership of defense-minded geeks at war with malware, effectively transforming himself into a digital Sam Spade—indeed, the graphic atop Shadowserver's home page features a Dashiell Hammett–style detective emerging from shadow.

Both sides in this cyberwar have become astonishingly sophisticated, operating at the cutting edge of programming theory and cryptography. Both understand the limits of security methodology, the one side working to broaden its reach, the other working to surpass it. Because malware has been automated, the good guys usually can only guess at who they are up against.

TROJANS, VIRUSES, AND WORMS

Rodney Joffe heads the cabal that has been battling Conficker. He is a burly, garrulous South African–born American who serves as senior vice president and chief technologist for Neustar, a company that provides trunk-line service for competing cell-phone companies around the world. Joffe's interest in stopping the worm did not stem just from his outrage and sense of justice. His concern for Neustar's operation is professional, and illustrative.

The company runs a huge local-number-portability database. Almost every phone call in North America, before it's completed, must ask Neustar where to go. Back in the old days, when the phone company was a monopoly, telecommunications were relatively simple. You could figure out where a phone call was going, right down to the building where the target phone would ring, just by looking at the number. Today we have competing telephone companies, and cell phones, and a person's telephone number is no longer necessarily tied to a geographic location. In this more complex world, someone needs to keep track of every single phone number, and know where to route calls so they end up in the right place. Neustar performs this service for telephone calls, and is one of many registries that oversee high-level Internet domains. It is, in Joffe's words, "the map."

"If I disappear, there's no map," he says. "So if you take us down, whole countries can actually disappear from the grid. They're connected, but no one can find their way there, because the map's disappeared."

A botnet like Conficker could theoretically be used to shut down Neustar's system. So Joffe helped form the Conficker Cabal. He scoffed when he read in late 2009 that the Obama administration's Department of Homeland Security planned to hire "a thousand" computer-security experts over the next three years. "There aren't more than a few hundred people in the world who understand this stuff."

Most of us use the word *virus* to describe all malware, but in geek-speak, it means something more specific. There are three types of the stuff: Trojans, viruses, and worms. A Trojan is a piece of software that works like a Trojan horse, masquerading as one thing to get inside a computer, and then attacking. A virus attacks the host computer after slipping in through a hole in its operating system. It depends on the computer-operator—you—doing something stupid to activate it, like opening an attachment to an e-mail that appears innocuous, or clicking on an enticing link. A worm works like a virus, exploiting flaws in operating systems, but it doesn't attack once it breaks in. It generally doesn't have a malicious payload. Exactly like the most-sophisticated viruses in the biological world, it does not cripple or kill its host. It is primarily designed to spread. The instructions that will put a worm like Conficker to work are not embedded in its code; they will be delivered later, from a remote command center.

In the old days, when your computer got infected, it slowed down because your commands had to compete for processing with viral invaders. You knew something was wrong because the machine took 10 times longer to boot up, or there was a delay between command and response. You began to get annoying pop-ups on your screen directing you to download supposedly remedial software. Programs would freeze. In this sense, the old malware was like the Ebola virus, a very scary strain that messily kills nearly everyone it infects— which is another way of saying that it is grossly ineffective, because it burns out the very host organisms it needs to survive. The miscreants who created computer viruses years ago learned that malware that announces itself in these ways doesn't last.

So today's malware produces no pop-ups, no slowdowns. A worm is especially quiet, since all it does, at least initially, is spread. Conficker stealthily sets up shop without making a ripple, and—other than calling home periodically for instructions—just waits. Its regular messages to its command center amount to only a couple hundred bytes of data, which is not enough to even light up the little bulb that flashes when a computer hard drive is at work.

After Phil Porras and others began snaring Conficker in increasing numbers, they began dissecting it. The worm itself was exquisite. It consisted of only a few hundred lines of code, no more than 35 kilobytes—slightly smaller than a 2,000-word document. In comparison, the average home computer today has anywhere from 40 to 200 *gigabytes* of storage. Unless you were looking for it, unless you knew *how* to look for it, you would never see it. Conficker drifts in like a mote.

It exploited a specific hole, Port 445, in the Microsoft operating systems, a vulnerability that the manufacturer had tried to repair just weeks earlier. Ports are designated "listening" points in a system, designed to transmit and receive particular kinds of data. There are many of them, more than 65,000, because an operating system consists of layer upon layer of functions. A firewall is a security program that guards these ports, controlling the flow of data in and out. Some ports, like the one that handles e-mail, are heavily trafficked. Most are not; they listen for updates and instructions that deal with a narrow and specific function, usually routine procedures that never rise to the notice of computer-users. Only certain very specific kinds of data can flow through ports, and then only with the appropriate codes. Windows opens Port 445 by default to perform tasks like issuing instructions for print-sharing or file-sharing. Late in the summer of 2008, Microsoft learned that even a system protected by a firewall was vulnerable at Port 445 if print-sharing and file-sharing were enabled (which they were on many computers). In other words, even a well-protected computer had a hole. On October 23, 2008, the company issued a rare "critical security bulletin" (MS08-067) with a patch to repair that hole. A specially crafted "remote procedure call" could allow the port to be used by a remote operator, the security bulletin warned, and "an attacker could exploit this vulnerability without authentication to run arbitrary code." The patch Microsoft offered theoretically slammed the door on a worm like Conficker almost a month before it appeared.

Theoretically.

In fact, the bulletin itself may have inspired the creation of Conficker. Many, many computer operators worldwide—you know who you are—fail to diligently heed security updates. And the patches are issued only to computers with validated software installations; millions of computers run on bootlegged operating systems, which have never been validated. Microsoft issues its updates on the second Tuesday of every month. Every geek in the world knows this; it's called "Patch Tuesday." The company employs some of the best programmers in the world to stay one step ahead of the bad guys. If everyone applied the new patches promptly, Windows would be nigh impregnable. But because so many people fail to apply the patches promptly, and because so many machines run on illegitimate Windows systems, Patch Tuesday has become part of Microsoft's problem. The company points out its own vulnerabilities, which is like a general responsible for defending a fort making a public announcement—"The back door to the supply shed in the southeast corner of the garrison has a broken lock; here's how to fix it." When there is only one fort, and it is well policed, the lock is fixed and the vulnerability disappears. But when you are defending millions of forts, and a goodly number of the people responsible for their security snooze right through Patch Tuesday, the security bulletin doesn't just invite attack, it provides a map! Twenty-eight days after the MS08-067 security bulletin appeared, Conficker started worming its way into unpatched computers.

THE CABAL'S SANDBOXES

Conficker's rate of replication got everyone's attention, so a loose-knit gaggle of geeky "good guys," including Porras, Joffe, and DiMino, began picking the worm apart. The online-security community consists of software manufacturers like Microsoft, companies like Symantec that sell security packages to computer owners, large telecommunication registries like Neustar and VeriSign, nonprofit research centers like SRI International, and botnet hunters like Shadowserver. In addition to maintaining honeypots, these security experts operate "sandboxes"—isolated computers (or, again, virtual computers inside larger ones) where they can place a piece of malware, turn it on, and watch it run. In other words, where they can play with it.

They all started playing with Conficker, comparing notes on what they found, and brainstorming ways to defeat it. That's when someone dubbed the group the "Conficker Cabal," and the name stuck, despite discomfort with the darker implications of the word. Here are some of the things the cabal discovered about the worm in those first few weeks:

* It patched the hole it came through at Port 445, making sure it would not have to compete with other worms. This was smart, because surely other hackers had seen security bulletin MS08-067.

* It tried to prevent communication with security providers (many computer-users subscribe to commercial services that regularly update antivirus software).

* When it started, if the IP address of the infected computer was Ukrainian, the worm self-destructed. When in attack mode, searching for other computers to infect, it skipped any with a Ukrainian IP address.

* It disabled the Windows "system restore" points, a useful tool that allows users with little expertise to simply reset an infected machine to a date prior to its infection. (System restore is one of the easiest ways to debug a machine.)

All of these things were clever. They indicated that Conficker's creator was up on all the latest tricks. But the main feature that intrigued the cabal was the way the worm called home. This is, of course, what worms designed to create botnets do. They settle in and periodically contact a command center to receive instructions. Botnet hunters like DiMino regularly wipe out whole malicious networks by deciphering the domain name of the command center and then getting it blocked. In the old days, this was easier because malware pointed to only a few IP addresses, which could be blocked by hosting providers and Internet service providers. The newer worms like Conficker bumped the game up to a higher level, generating domain names that involve many providers and a wide range of IP addresses, and that security experts can block only by contacting Internet registries—organizations that manage the domain registrations for their realm. But Conficker did not call home to a fixed address.

Shortly after it was discovered, the worm began performing a new operation: generating a list of domain names seemingly at random, 250 a day across five top-level domains (top-level domains are defined by the final letters in a Web address, such as *.com* or *.edu* or *.uk*). The worm would then go down the list until it hit upon the one connected to its remote controller's server. All Conficker's controller had to do was register one of the addresses, which can be done for a fee of about $10, and await the worm's regular calls. If he wished, he could issue instructions. It was as if the boss of a crime family told his henchmen to check in daily by turning to the bottom of a certain page in each day's *Racing Form,* where there would be a list of potential numbers. They would then call each number until the boss picked up. So it was not apparent from day to day where the worm would call home.

With the *Racing Form* trick, if you were a cop and were tipped off where to look, you might arrange with the paper's publisher to see the page before it was printed, and thus be one step ahead of the henchmen and their boss. To defeat Conficker, the geeks would have to figure out in advance what the numbers (or, in this case, domain names) would be, and then hustle to either buy up or contact every one, block it, or cajole whoever owned it to cooperate before the worm "made the call."

Michael Ligh, a young Brooklyn researcher employed by the computer-security company iDefense, is one of several people who went to work unraveling Conficker's methods. Ligh and others had seen algorithms for random-domain-name generation before, and most were keyed to the infected computer's clock. If new places to call home must be generated every day, or every few hours, then the worm needs to know when to perform the procedure. So the malware simply checks the time on its host computer. This provided the good guys with a tool to defeat it. They turned the clock forward on their sandbox computer, forcing their captured strain of the worm to spit out all the domain names it would generate for as long into the future as they cared to look. It was like stealing the teacher's edition of a classroom textbook, the one with all the answers to the quizzes and tests printed in the back. Once you knew all the places the malware would be calling, you could cordon off those sites in advance, effectively stranding the worm.

Conficker had an answer for that. Instead of using the infected computer's clock, the worm set its schedule by the time on popular corporate home pages, like Yahoo, Google, or Microsoft's own msn. com.

"*That* was interesting," Ligh said. "There was no way we could turn the clock forward on Google's home page."

So there was no easy way to predict the list of domain names in advance. But there was *a* way. The first step was to set up a proxy server to, in effect, intercept the time update from the big corporate Web site before it got back to the worm, alter the information, and then send it on. You could then tell the worm it was a date sometime in the future, and the worm would spit out the domain names for that date. This was a tedious way to proceed, since you could generate only one set of new domain names at a time. So Ligh and other researchers reverse-engineered the worm's algorithm, extracted the time-update function, and wedded it to a piece of code they could control. They instructed their copy to generate the future lists in advance. They could then buy up or block all the sites, and direct all the worm's communications into a "sinkhole," a dead-end location where calls go unanswered. Conficker's creators had deliberately made the task so onerous and expensive that *no one* would go to the trouble of blocking all possible command centers.

Or so they thought. The cabal, through a determined and unprecedented effort, did manage to cordon off the worm. By the end of 2008, Conficker had infected an estimated 1.5 million machines worldwide, but it was on its way to full containment. In the great chess match, the good guys had called "Check!"

Then the worm turned.

MD-6

On December 29, 2008, a new version of Conficker showed up, and if the geeks had been intrigued with the original version, they now experienced something more akin to respect . . . mingled with fear.

One of the early theories about the worm was that it had slipped out of a computer-science lab, the product of some fooling-around by a sophisticated graduate student or group of students. They had loosed it on the world inadvertently, or maybe on purpose as a prank or experiment without realizing how effective it would be. This hypothesis appealed to optimists.

The new version of the worm, Conficker B, exploded the benevolent-accident theory. It was clear that the worm's creator had been watching every move the good guys made, and was adjusting accordingly. He didn't care that the good guys could predict its upcoming lists of domain names. He just rejiggered the worm to spread the new lists out over eight top-level domains instead of five, making the job of blocking them far more difficult. The worm had no trouble contacting all of these locations. If it received no command from one, it simply tried the next one on its list. Conficker B could go on like this for months, even years. It had to find its controller only once to receive instructions.

"That's a high number," Rodney Joffe, of Neustar, told me. "The cops will get sick and tired of knocking on 250 doors a day and finding there's no one there. And if I'm the chief bad guy, all I have to do is be behind one of those doors on one of those days."

There were other improvements to Conficker. Among them: besides shutting down whatever security system was installed on the computer it invaded, and preventing it from communicating with computer-security Web sites, it stopped the computer from connecting with Microsoft to perform Windows updates. So even though Microsoft was providing patches, the infected machines could not get to them. In addition, it modified the computer's bandwidth settings to increase speed and propagate itself faster; and it began to spread itself in different ways, including via USB drives. This last innovation meant that even "closed" computer networks, those with no connection to the Internet, were vulnerable, since users who cannot readily transmit files from point to point via the Web often store and transport them on small USB drives. If one of those USB drives, or a CD, was plugged into an infected computer, it could deliver the worm to an entire closed network.

All of this was impressive—but something else stopped researchers cold. Analysts with Conficker B isolated in their sandboxes could watch it regularly call home and receive a return message. The exchange was in code, and not just any code.

Breaking codes used to be the province of clever puzzle masters, who during World War II devised encryption and code-breaking methods so difficult that operators needed machines to do the work. Computers today can perform so many calculations so fast that, theoretically at least, no cipher is too difficult to crack. One simply applies what computer scientists call "brute force": trying every possible combination systematically until the secret is revealed. The game is to make a cipher so difficult that the amount of computing power needed to break it renders the effort pointless—the "thief" would have to spend more to obtain the prize than the prize is worth. In his 1999 history of code-making and -breaking, *The Code Book*, Simon Singh wrote: "It is now routine to encrypt a message [so securely] that all the computers on the planet would need longer than the age of the universe to break the cipher."

The basis for the highest-level modern ciphers is a public-key encryption method invented in 1977 by three researchers at MIT: Ron Rivest (the primary author), Adi Shamir, and Leonard Adleman. In the more than 30 years since it was devised, the method has been improved several times. The National Institute of Standards and Technology sets the Federal Information Processing Standard, which defines the cryptography algorithms that government agencies must use to protect communications. Because it is the most sophisticated oversight effort of its kind, the standard is determined by an international competition among the world's top cryptologists, with the winning entry becoming by default the worldwide standard. The current highest-level standard is labeled SHA-2 (Secure Hash Algorithm–2). Both this and the first SHA standard are versions of Rivest's method. The international competition to upgrade SHA-2 has been under way for several years and is tentatively scheduled to conclude in 2013, at which point the new standard will become SHA-3.

Rivest's proposal for the new standard, MD-6 (Message Digest–6), was submitted in the fall of 2008, about a month before Conficker first appeared, and began undergoing rigorous peer review—the very small community of high-level cryptographers worldwide began testing it for flaws.

Needless to say, this is a very arcane game. The entries are comprehensible to very few people. According to Rodney Joffe, "Unless you're a subject-matter expert actively involved in crypto-algorithms, you didn't even know that MD-6 existed. It wasn't like it was put in *The New York Times*."

So when the new version of Conficker appeared, and its new method of encrypting its communication employed MD-6, Rivest's *proposal* for SHA-3, the cabal's collective mind was blown.

"It was clear that these guys were not your average high-school kids or hackers or predominantly lazy," Joffe told me. "They were making use of some very, very sophisticated techniques.

"Not only are we not dealing with amateurs, we are possibly dealing with people who are superior to all of our skills in crypto," he said. "If there's a surgeon out there who's the world's foremost expert on treating retinitis pigmentosa, he doesn't do bunions. The guy who is the world expert on bunions—and, let's say, bunions on the third digit of Anglo-American males between the ages of 35 and 40, that are different than anything else—he doesn't do surgery for retinitis pigmentosa. The knowledge it took to employ Rivest's proposal for SHA-3 demonstrated a similarly high level of specialization. We found an equivalent of three or four of those in the code—different parts of it.

"Take Windows," he explained. "The understanding of Windows' operating system, and how it worked in the kernel, needed that kind of a domain expert, and they had that kind of ability there. And we realized as a community that we were not dealing with something normal. We're dealing with one of two things: either we're dealing with incredibly sophisticated cyber criminals, or we're dealing with a group that was funded by a nation-state. Because this wasn't the kind of team that you could just assemble by getting your five buddies who play Xbox 360 and saying, 'Let's all work together and see what we can do.'"

The plot thickened—it turned out that Rivest's proposal, MD-6, had a flaw. Cryptologists in the competition had duly gone to work trying to crack the code, and one had succeeded. In early 2009, Rivest quietly withdrew his proposal, corrected it, and resubmitted it. This gave the cabal an opening. If the original Rivest proposal was flawed, then so was the encryption method for Conficker B. If they were able to eavesdrop on communications between Conficker and its mysterious controller, they might be able to figure out who he was, or who they were. How likely was it that the creator of Conficker would know about the flaw discovered in MD-6?

Once again, the good guys had the bad guys in check.

About six weeks later, another new version of the worm appeared.

It employed Rivest's *revised* MD-6 proposal.

Game on.

"Our Finest Hour"

By early 2009, Conficker B had infected millions of machines. It had invaded the United Kingdom's Defense Ministry. As CBS prepared a

60 Minutes segment on the worm, its computers were struck. In both instances, security experts scrambled to uproot the invader, badly disrupting normal functioning of the system. Conficker now had the world's attention. In February 2009, the cabal became more formal. Headed initially by a Microsoft program manager, and eventually by Joffe, it became the Conficker Working Group. Microsoft offered a $250,000 bounty for the arrest and conviction of the worm's creators.

The newly named team went to work trying to corral Conficker B. Getting rid of it was out of the question. Even though they could scrub it from an infected computer, there was no way they could scrub it from all infected computers. The millions of machines in the botnet were spread all over the world, and most users of infected ones didn't even know it. It was theoretically feasible to unleash a counter-worm, something to surreptitiously enter computers and take out Conficker, but in free countries, privacy laws frown on invading people's home computers. Even if all the governments got together to allow a massive attack on Conficker—an unlikely event—the new version of the worm had new ways of evading the threat.

Conficker C appeared in March 2009, and in addition to being impressed by its very snazzy crypto, the Conficker Working Group noticed that the new worm's code threatened to up the number of domain names generated every day to 50,000. The new version would begin generating that many domain names daily on April 1. At the same time, all computers infected with the old variants of Conficker that could be reached would be updated with this new strain. The move suggested that the bad guys behind Conficker understood not just cryptology, but also the mostly volunteer nature of the cabal.

"You know you're dealing with someone who not only knows how botnets work, but who understands how the security community works," Andre' DiMino told me. "This is not just a bunch of organized criminals that, say, commission someone to write a botnet for them. They know the challenges that the security community faces internally, politically, and economically, and are exploiting them as well."

The bad guys knew, for instance, that preregistering even 250 domain names a day at $10 a pop was doable for the good guys. As long as the number remained relatively small, the cabal could stay ahead of them. But how could the good guys cope with a daily flood of 50,000? It would require an unprecedented degree of cooperation among competing security firms, software manufacturers, non-profit organizations like Shadowserver, academics, and law enforcement.

"You can't just register all 50,000—you've got to go one by one and make sure the domain name doesn't already exist," Joffe says. "And if it exists, you've got to make sure that it belongs to a good guy, not a bad guy. You've got to make a damn phone call for any of the new ones, and have to send someone out there to do it—and these are spread all over the world, including some very remote places, Third World countries. Now the bar had been raised to a level that was almost insurmountable."

The worm was already running rings around the good guys, and then, just for good measure, it planted a pie in their faces on, of all days, April 1. By playing with the new variant in their sandboxes, the cabal knew that the enhanced domain-name-generating algorithm would click in on that day. If the update succeeded, it would be a game-changer. It was the most dramatic moment since Conficker had surfaced the previous November. Apparently, at long last, this extraordinary tool was going to be put to use. But for what? The potential was scary. Few people outside the upper echelon of computer security even understood what Conficker was, much less what was at stake on April 1, but word of a vague impending digital doomsday spread. The popular press got hold of it. There were headlines and the usual spate of ill-informed reports on cable TV and the Internet.

When the day arrived, those who had been warning about the dangers of this new worm were sure to see their fears vindicated.

The cabal mounted a heroic effort to shut down the worm's potential command centers in advance of the update, coordinating directly with the Internet Corporation for Assigned Names and Numbers, the organization that supervises registries worldwide. "It was our finest hour," Joffe says.

"I don't think that the bad guys could have expected the research community to come together as it did, because it was pretty unprecedented," Ramses Martinez, director of information security for VeriSign, told me. "That was a new thing that happened. I mean, if you would have told me everybody's going to come together—by *everybody*, I mean all these guys in this computer-security world that know each other—and they're going to do this thing, I would have said, 'You're crazy.' I don't think the bad guys could have expected that."

Much of the computer world was watching, in considerable suspense, to see what would happen on April 1. It was like the moment in a movie when the bad guy at last has cornered the hero. He pulls out an enormous gun and aims it at the hero's head, pulls the trigger . . . and out pops a little flag with the word BANG!

Conficker found one or two domain names that Joffe's group had missed, which was all it needed. The cabal's efforts had succeeded in vastly reducing the number of machines that got the update, but the ones that did went to work distributing a very conventional, well-known malware called Waledac, which sends out e-mail spam selling a fake anti-spyware program. The worm was used to distribute Waledac for two weeks, and then stopped.

But something much more important had happened. The updated worm didn't just up the ante by generating 50,000 domain names daily; it effectively moved the game out of the cabal's reach.

* * *

"APRIL 1 CAME AND WENT, and in the middle of that night the systems switched over to the new algorithm," Conficker C, Joffe told me. "That's all that was supposed to happen, and it happened. But the Internet didn't get infected; it was just an algorithm change in the software. So of course the press said, 'Conficker is a bust.'"

Public concern over the worm fizzled, just as the problem grew worse: the new version of Conficker introduced peer-to-peer communications, which was disheartening to the good guys, to say the least. Peer-to-peer operations meant the worm no longer had to sneak in through Windows Port 445 or a USB drive; an infected computer spread the worm directly to every machine it interacted with. It also meant that Conficker no longer needed to call out to a command center for instructions; they could be distributed directly, computer to computer. And since the worm no longer needed to call home, there was no longer any way to tell how many computers were infected.

In the great chess match, the worm had just pronounced "Checkmate."

WATCHING AND WAITING

As of this writing, 17 months after it appeared and about a year after the April 1 update, Conficker has created a stable botnet. It consists of anywhere from hundreds of thousands of computers to 12 million. No one knows for sure anymore, because with peer-to-peer communications, the worm no longer needs to check in with an outside command center, which is how the good guys kept count. Joffe estimates that with the four distinct strains (yet another one appeared on April 8, 2009), 6.5 million computers are probably infected.

The investigators see no immediate chance or even any effective way to kill it.

"There are a bunch of infected machines that are out there, and they can be taken over, given the right circumstances, by the bad guys," VeriSign's Martinez says. "Will they do that? I don't know. So it's a potential threat. It's something that's out there, sitting there, and it needs to be addressed, but I don't think, honestly, that we know how. How do we address this? If it was sitting in the U.S., it would be a fairly easy thing to do. The fact is that it's spread out all around the world."

Ever since the paltry Waledac scam, the worm has been biding its time.

"They are watching us watch them," says Andre' DiMino, the botnet hunter. "I think it's really either that or somebody let this thing get bigger, and it's advanced bigger and further than they ever dreamed possible. A lot of people think that. But in looking at the sophistication of this thing and looking at the evolution of this thing, I think they knew exactly what they were doing. I think they were trying something, and I think that they're too smart to do what everybody figured they were going to do. You have to remember, the world was watching this thing and waiting for the world to end from Conficker on April 1, 2009. The last thing you'd want to do if you're the bad guy is make something happen on April 1. You're never going to do that, because everybody's watching it. You're going to do something when you're least suspected. So these guys are sophisticated. They have good code. And just even seeing the evolution from Conficker A to B to C, where there's the peer-to-peer component, which . . . strikes fear into the heart of botnet hunters because it's just so damn difficult to track—these guys know exactly what they're doing."

SO WHO ARE THEY?

One of the things Martinez's team does, patrolling the perimeter at VeriSign looking for threats, is dip into the obscure digital forums where cyber criminals converse. Those who are engaged in writing sophisticated malware boast and threaten and compare notes. The good guys venture in to collect intelligence, or just out of curiosity, or for fun. They sometimes pretend to be malware creators themselves, sometimes not. Sometimes they engage in a little cyber trash talk.

"In the past you were just sort of making sure they didn't steal your proprietary information," Martinez says. "Now we go in to engage them. You talk to them and you exchange information. You have a guy in Russia selling malware, working with a guy in Mexico doing phishing attacks, who's talking to a kid in Brazil, who's doing credit-card fraud, and they're introducing each other to some guy in China doing something else."

Martinez said he recently eavesdropped on a dialogue between a security researcher and a man he suspects was at least partly responsible for Conficker. He wouldn't say how he drew that connection, only that he had good reasons for believing it to be true. The suspect in the conversation was eastern European. The standard image of a malware creator is the Hollywood one: a brilliant 20-something with long hair and a bad attitude, in need of a bath. This is not how Martinez sees his nemesis—or nemeses.

"I see him, or them, as a really well-educated, smart business-man," he said. "He may be 50 years old. These guys are not chumps. They're not just out to make a buck."

The eastern European, backpedaling from further dialogue with the security geek, wrote, "You're the good guys; we're the bad guys. Bacillus can't live with antibodies."

"Now, I didn't grow up in a bad neighborhood or anything," said Martinez, "but the few thugs that I saw would never use a word like *bacillus* or make an analogy like that."

One of the early clues in the hunt was the peculiarity in the Conficker code that made computers with active Ukrainian keyboards immune. Much of the world's aggressive malware comes from eastern

Europe, where there are high levels of education and technical exper- tise, and also thriving organized criminal gangs. Martinez believes Conficker was written by a group of highly skilled programmers. Like Joffe, he sees it as a group of creators, because designing the worm re- quired expertise in so many different disciplines. He suspects that these skilled programmers and technicians either were hired by a crim- inal gang, or created the worm as their own illicit business venture. If that's true, then the Waledac maneuver was like flexing Conficker's pinkie—just a demonstration, a way of showing that despite the best and most concerted effort of the world's computer-security establish- ment, the worm was fully operational and under their control.

Will they be caught?

"I have no idea," Martinez says. "I would say probably not. I'll be shocked if they're ever arrested. And arrest them for what? Is break- ing into people's computers even illegal where they're from? Because in a lot of countries, it isn't. As a matter of fact, in some countries, unless you're touching a computer in their jurisdiction, their coun- try, that's not illegal. So who's going to arrest them, even if we know who they are?"

Ridding computers of the worm poses another kind of over- whelming problem.

"THERE ARE CONTROLS, OR CHECKS and balances, in place to limit what police can do, because we have civil liberties to protect," he says. "If you do away with these checks and balances, where the government can come in and reimage your computer overnight, now you're infringing on people's civil liberties. So, I mean, we can talk about this all day, but I'll tell you, it's going to be a long time, in my opinion, before we really see the government being able to effec- tively deal with cyber crime, because I think we're still learning as a culture, as a nation, and as a world how to deal with this stuff. It's too new."

Imagining Conficker's creators as a skilled group of illicit cyber entrepreneurs remains the prevailing theory. Some of the good guys feel that the worm will never be used again. They argue that it has become too notorious, too visible, to be useful. Its creators have learned how to whip computer-security systems worldwide, and will now use that knowledge to craft an even stealthier worm, and perhaps sell it to the highest bidder. Few believe Conficker itself is the work of any one nation, because other than the initial quirk of the Ukrainian-keyboard exemption, it spreads indiscriminately. China is the nation most often suspected in cyber attacks, but there may be more Conficker-infected computers in China than anywhere else. Besides, a nation seeking to create a botnet weapon is unlikely to create one as brazen as Conficker, which from the start has exhibited a thumb-in-your-eye, catch-me-if-you-can personality. It is hard to imagine Conficker's creators not enjoying the high level of cyber gamesmanship. The good guys certainly have.

"It's cops and robbers, so to speak, and that was a really interesting aspect of the work for me," says Martinez. "It's guys trying to outwit each other and exploit vulnerabilities in this vast network."

In chess, when your opponent checkmates you, you have no recourse. You concede and shake the victor's hand. In the real-world chess match over Conficker, the good guys have another recourse. They can, in effect, upend the board and go after the bad guys physically. Which is where things stand. The hunt for the mastermind (or masterminds) behind the worm is ongoing.

"It's an active investigation," Joffe says. "That's all I can say. Law enforcement is fully engaged. We have some leads. This story is not over."

PETER J. BOYER

The Covenant

FROM *THE NEW YORKER*

*Because he is an avowed Christian, some feel that Francis Collins,
director of the National Institutes of Health, should not be in a
position where his beliefs could play a role in decisions about science. Peter J. Boyer profiles this controversial figure.*

WHEN THE GENETICIST FRANCIS COLLINS WAS
named director of the National Institutes of Health, last
summer, he became the public face of American science
and the keeper of the world's deepest biomedical-research-funding
purse. He was praised by President Obama and waved through the
Senate confirmation process without objection. There also came a
peer review of a sort that he'd never experienced, conducted in the
press and in Internet science forums. Collins read in the *Times* that

many of his colleagues in the scientific community believed that he suffered from "dementia." Steven Pinker, a cognitive psychologist at Harvard, questioned the appointment on the ground that Collins was "an advocate of profoundly anti-scientific beliefs." P. Z. Myers, a biologist at the University of Minnesota at Morris, complained, "I don't want American science to be represented by a clown."

Collins's detractors did not question his professional achievements, which long ago secured his place in the first rank of international scientists. As a young researcher at Yale, Collins conceived a method of hastening the laborious process of hunting disease-causing genes by skipping across long stretches of chromosomes until the suspect gene's neighborhood was located. As an assistant professor at the University of Michigan, in the nineteen-eighties, he and collaborators at the University of Toronto employed this method to find the gene that causes cystic fibrosis and, a year later, the genetic flaw responsible for neurofibromatosis. These breakthroughs brought him fame and, eventually, the job of director of the Human Genome Project, which promised to revolutionize medicine by identifying and mapping all the approximately twenty thousand human genes that code for protein.

Thanks to that job, there wasn't much doubt about Collins's ability to handle the formidable management challenge of running the N.I.H., which directly employs twenty thousand scientists and staff, funds three hundred and twenty-five thousand outside researchers, and operates twenty-seven institutes and research centers on its campus, in Bethesda, Maryland. A key duty of the N.I.H. director is to justify the agency's budget and defend before Congress the programs it funds, a duty that requires a skill quite apart from prowess in the laboratory. In fifteen years at the National Human Genome Research Institute, Collins had proved himself an able manager, bringing the Genome Project to a successful conclusion in 2003— two and a half years early and four hundred million dollars under budget. He also won friends in Congress with a genial manner and a gift for conveying complex scientific information in felicitous language.

The objection to Collins was his faith—or, at least, the ardency of it. Collins is a believing Christian, which places him in the minority among his peers in the National Academy of Science. (Of its members, according to a study, only seven per cent believe in God.) After leaving the Genome Research Institute, Collins began drawing large crowds on the college lecture circuit; he created a Web site, BioLogos, to advance his idea of the companionability of reason and faith; and he wrote a best-selling book, *The Language of God,* in which he presented what he claims to be scientific evidence of the existence of God.

President Obama's choice of Collins for the N.I.H. touched a nerve. The George W. Bush era had been an extraordinarily fractious time in public science, beginning with Bush's first prime-time address to the nation, in which he announced restrictions on embryonic-stem-cell research. That move, and others that followed, convinced Bush's critics that the religious right had become the final arbiter of public policy, an impression that Bush seemed little inclined to dispel. "Well, we thought we'd seen the last of the theocracy of George W. Bush, but it apparently ain't so," Dr. Jerry Coyne, a University of Chicago professor, wrote when Collins was appointed. "I am funded by the N.I.H., and I'm worried. Not about my own funding (although I'm a heathen cultural Jew), but about how this will affect things like stem-cell research and its funding."

A year later, Obama's appointment of Collins seemed an inspired choice. The President had found not only a man who reflected his own view of the harmony between science and faith but an evangelical Christian who hoped that the government's expansion of embryonic-stem-cell research might bring the culture war over science to a quiet end. On August 23rd, however, Judge Royce C. Lamberth, of the Federal District Court for the District of Columbia, halted federal spending for embryonic-stem-cell research, putting hundreds of research projects in limbo and plunging the N.I.H. back into a newly contentious national debate.

AT THE N.I.H., THE ABILITY to deal with controversies, as a generation of Collins's predecessors learned, matters at least as much as credentials; political combat comes with the job. Collins does not seem a likely combatant. His physical aspect—gray mustache and hair (cut in an early-Beatles mop top), thin-rimmed eyeglasses, and a distinct pallor—suggests a man best acquainted with a sunless existence in some laboratory. Yet, in a relatively colorless town, Collins has come to be known as something of a character, a model of geek cool. He likes big, noisy motorcycles, and, despite a mild manner, he is famously unself-conscious. At the unlikeliest moments, he will strap on a guitar and accompany himself in song, often a tune he has composed for the occasion.

In dealing with Congress, Collins is less given to sentiment. A few weeks after moving into the director's office, he received a letter from two Republican congressmen pressing him about a handful of N.I.H. grants "that do not seem to be of the highest scientific rigor." The lawmakers, Joe Barton, of Texas, and Greg Walden, of Oregon, demanded to know the process by which the agency had funded a $423,500 study of why young heterosexual men did not consistently use condoms during sex. Barton and Walden were also curious about an N.I.H.-funded study investigating substance use and H.I.V.-risk behaviors among female and transgender sex workers in Thailand, and a $29,469 grant to researchers studying patterns of drug abuse in the Brazilian rave culture. The Barton-Walden inquiry was, in practical terms, just a gesture, as the programs in question had already been funded. Sometimes, though, such a gesture hits a vein of real political opportunity.

In November, Barton and Walden sent Collins a second letter, this time focussing on a series of N.I.H. grants investigating gun violence as a health issue. The N.I.H.-sponsored study, if portrayed as a back-door attempt at gun control (as it immediately was by Glenn Beck, on Fox News), threatened an unwelcome contretemps for an Administration fast depleting its political capital.

WHEN I ASKED COLLINS ABOUT the Barton-Walden inquiries last winter, his response was polite but dismissive. "These grants, like all of our grants, go through this rigorous, two-level peer-review process," he said. "And you can go back with any grant, and you can see, O.K., who was the peer-review group that looked at this? O.K., these are the leaders of their field. And what kind of evaluation did they give it? Well, it got funded, which means it was in the top twenty per cent of grants that that group looked at." That was the message Collins conveyed to Barton and Walden in his written response to their queries, and the issue faded away.

The Barton-Walden exchange reflects a paradox faced by any director of the N.I.H.: the public (and, therefore, Congress) likes the idea of spending tax money on medical discoveries, but often recoils when presented with details of the research (especially research outside the laboratory) that it's bought. The N.I.H. was a politically obvious choice when the Administration and congressional Democrats were dispersing stimulus money, and the Institutes received ten billion dollars to spend over two years. But, inevitably, some of that money went toward projects that made easy targets for Republicans. This summer, Senators John McCain and Tom Coburn issued a list of "stimulus projects that give taxpayers the blues," and the N.I.H. found itself under attack for a $180,935 grant to the University of Missouri to find better ways to freeze rat sperm.

"Now, why would anybody spend money on freezing rat semen?" Collins asks. "Well, I'll tell you why. We have all these incredibly valuable rat strains that represent particular models of human disease, like hypertension or heart disease. But you don't necessarily want to keep them running around in cages gobbling up rat food at extreme expense, year after year, if you're not sure that strain is something you're going to want to study five years from now. If you just freeze the sperm, you can re-create that rat when you're ready, and it saves us a huge amount of money. Knowing how to do that effectively is a pretty good investment. But, of course, nobody bothered to find out the reason for this. They just thought it sounded weird and bizarre and like a waste of money."

THE MAN WHO HOLDS THE most powerful job in American science came from an unusual background. During the Depression, Collins's parents, Fletcher and Margaret Collins, became part of a short-lived West Virginia project—sponsored by Eleanor Roosevelt, and with financial help from Bernard Baruch—that attempted to create an ideal community for a group of impoverished miners near Morgantown.

Fletcher was the project's music director, with the mission of helping the homesteaders recover their cultural heritage. He had a gift for coaxing from the mountain people the nearly forgotten old fiddle tunes, folk songs, and square-dance calls that had been, he wrote, "very much in their blood," but "layered over by coal dust."

After the war, the Collinses bought a ninety-four-acre farm in the Shenandoah Valley, near Staunton, Virginia, determined to derive a livelihood from the land without modern agricultural machinery. They kept chickens, sheep, cows, and two workhorses, who pulled the plow and old wagon that carried the harvest from the hilly fields to the barn. The four Collins children, all boys, served as farmhands, collecting eggs, milking the cows, and shucking corn. When the alfalfa needed to be mowed and baled, amused neighbors would stop by with their tractors and help out. Once a week, the family drove into Staunton, where Margaret's parents lived, and the boys had a bath; in the summertime, they bathed in the cow trough. After a few years, Fletcher took a position as drama instructor at the local women's college, Mary Baldwin ("My cash crop," he'd say), but the family was relatively poor. The younger boys wore their brothers' hand-me-downs, and by the time the clothes reached Francis, the youngest, they were threadbare.

The Collinses' household, known as Pennyroyal Farm, became the center of a vibrant arts community in Staunton. (It's still thriving.) "Musicians would come and crash there for a couple of weeks because they'd run out of money," Collins recalls. "They'd play great music, and then finally they'd move on." Bob Dylan was among those who came to Pennyroyal. "Margaret and Fletcher were sort of hippies before there were hippies," the singer Linda Williams recalls. "They were back-to-the-landers, and saw things the way people did in the seventies, only they'd done it in the thirties."

For Francis, it was an enchanting, if arduous, childhood, part *Boys' Life* and part Woodstock. He could set a barn door and knew how to predict weather by reading the sky over the distant Alleghenies. He did not see the inside of a schoolroom until sixth grade, because Margaret taught her boys at home. "There was no schedule," Francis recalls. "The idea of Mother having a lesson plan would be just completely laughable. But she would get us excited about trying to learn about a topic that we didn't know much about. And she would pose a question and basically charge you with it, using whatever resources you had—your mind, exploring nature, reading books—to try to figure out, well, what could you learn about that? And you'd keep at it until it just got tiresome. And then she'd always be ready for the next thing."

Collins doesn't recall ever seeing a Bible among the books in his parents' house, and the only time they sent him to church it was for the choir music, and with the admonition "You should be respectful of what they're doing, even if the stuff they're talking about doesn't make a lot of sense." He was an agnostic when he went off to the University of Virginia, and by the time he was studying physical chemistry as a graduate student, at Yale, he'd become what he calls a "fundamentalist" atheist—the sort of non-believer who would share his dining table with a believer, just for the chance to expose the folly of faith. "I was fairly obnoxious about it," he says.

Collins's academic career was a sprint. He published his first research paper while still an undergraduate at Virginia. The summer after his junior year, he married Mary Lynn Harman, a struggling math student he'd tutored in high school, and they soon had two children. (They have since divorced, and Collins has remarried.) Collins was breezing through a doctoral program in theoretical chemical physics at Yale when he realized that a professor's life in that field was not at all what he wanted. He decided instead to become a physician, and enrolled at the University of North Carolina School of Medicine.

Collins loved medical school, but he worked impossible hours; on weekends he was finishing his physics dissertation. He and Mary Lynn moved to a small working-class town called Carrboro, at the edge of Chapel Hill, and she began to attend services at a little Methodist tabernacle. Eventually, her faith caught fire. She hoped that Francis would try the church, too, but he emphatically declined, telling her he didn't wish to hear any more "about this Jesus junk."

Collins had been feeling a nagging of a different sort. As a physician in training, he frequently found himself at the bedside of desperately ill patients, many of whom displayed a surprising equanimity and were only too happy to tell him why. He heard countless stories of faith, mostly of the Protestant Christian variety predominant in the South, and he noted the power of a psychological crutch, firmly held. One day, an elderly woman suffering from untreatable acute angina asked Collins what he believed.

He had nothing to say, which slightly embarrassed him; he was more bothered by the realization that he didn't know why he didn't believe. He decided that the question of whether God existed was an important one, and undertook to affirm his atheism. He spent two years reading C. S. Lewis and other explorations of reason and faith, and eventually called Mary Lynn's pastor, the Reverend Sam McMillan.

McMillan recalls, "He said, 'Please don't be offended, but I'd have to get a lobotomy to go to your church.'" McMillan played golf, and Collins met him at the course for a round. Collins assailed the young pastor with questions.

"I'd be just about to putt a key putt, and he'd say, 'You don't believe in the Virgin Mary, do you?'" McMillan says. "By the eighteenth hole, I knew we were doing more than playing golf. And somehow I was prompted to seize the moment. It was the Holy Spirit inspiration. I grabbed the scorecard and wrote down, 'When God knocks on my door, in a way that I—not my wife or pastor, but I— know that it's God who's knocking on my door, I will then accept Jesus Christ.' I gave it to him. And he signed it." (McMillan, who runs a Christian retreat in Boone, North Carolina, still uses that ploy with potential converts; he calls it the Francis Collins Covenant.) McMillan did not hear from Collins for several months. Then Collins called him while hiking in the Cascades, and showed up at church the next Sunday, saying he wanted to share his testimony. He told the congregation that during his trip he turned a corner and saw a frozen waterfall, perfectly formed into three separate parts. He took it as a revelation of Trinitarian truth, the sign that he'd contracted for on Sam McMillan's golf card. The next morning, he vowed to devote his life to the Christian faith.

LAST SUMMER, SOME OF THE world's preëminent scientists were in Cambridge, England, celebrating the hundred-and-fiftieth anniversary of the publication of *On the Origin of Species*, when word arrived from Washington that Obama had chosen Collins to run the N.I.H. "A lot of those New Atheists were in attendance, and they were incredibly alarmed," Harold Varmus, a Nobel laureate, who was Bill Clinton's director of the N.I.H., says. Varmus, who now runs the National Cancer Institute, is an old friend of Collins's, and was his boss during the Human Genome Project. "I tried to calm everybody down," he says. "I said, Francis is a terrific scientist, and very well organized and a great spokesperson for the N.I.H., has terrific connections in Congress, and is a delightful person to work with."

It was clear that Collins's handling of stem-cell policy would be the critical test of his ability to separate faith from secular duty. James Battey, the vice-chairman of the stem-cell task force at the N.I.H., says that the issues surrounding stem-cell research were far more nuanced and difficult than the Bush-era political argument made them seem. Stem cells were portrayed as holding an almost magical promise for the treatment of disease, and with good reason. The cells, derived from a human embryo at the blastocyst stage (five days after its formation), can renew themselves indefinitely, giving scientists a virtually unlimited supply to experiment with. And they have the capacity to become any cell type in the body. They are called pluripotent cells, for their remarkable variability. "If you want a cell line that you can somehow differentiate into, say, dopamine-producing neurons, to help a patient with Parkinson's disease, or insulin-producing cells, to help a patient with diabetes, these cells clearly have the capacity to do that," Battey says. That is an enticing prospect, but so far no treatments have resulted from embryonic-stem-cell research.

The principal ethical issue posed by the research—the fact that the embryo must be destroyed in the process—was a cause of real concern even to the man who pioneered the process, Dr. James Thomson, of the University of Wisconsin. "If human-embryonic-stem-cell research does not make you at least a little bit uncomfortable, you have not thought about it enough," he said in 2007.

The political problem surrounding such research arose from the fact that in 1995 Congress prohibited federal money from being used in any research that destroys an embryo. This action, called the

Dickey-Wicker amendment (named after its Gingrich-era Republican authors, Jay Dickey and Roger Wicker), echoed the tactics of pro-life forces in the abortion wars. Although many moderate Republicans and pro-life Democrats supported promising new biomedical research, such support became problematic when framed as a means of killing embryos. Dickey-Wicker was a rider attached to the appropriations bill for the Department of Health and Human Services (N.I.H.'s parent agency). It was signed by President Clinton, and has been passed, and signed, every year since.

In 1999, the Clinton Administration devised a way of getting around the restriction, with a legal finding that federal money could be used on the research if the process of extracting the cells was paid for with private funds. Clinton was out of office before that policy could be tried, which put the matter on the desk of George W. Bush. The early months of Bush's presidency featured an ever-heightening debate over embryonic-stem-cell research, which is why it became the subject of his first television address, in the summer of 2001.

Bush's solution was to allow funding for research using twenty-one stem-cell lines that already existed as of the date that he announced his policy—August 9, 2001. The Bush policy also restricted federally funded research to stem cells that were derived from embryos that were no longer needed for reproductive purposes; and it mandated informed consent from the embryos' donors. In the meantime, private laboratories were free to create and experiment on pluripotent cells without restriction. "For that matter, you could even engage in human reproductive cloning in this country, if you wanted to," Battey says. "It isn't prohibited in some states."

There things stood until March of last year, when Obama announced a new policy, formulated partly on the advice of Francis Collins. In the years since Bush's announcement, there had been a remarkable development in stem-cell research—one that offered Obama and Collins an out, had they chosen to take it. Partly because of the Bush restrictions, James Thomson and Shinya Yamanaka, a pioneer in pluripotent-cell research at Kyoto University, developed a method for creating cells that mimicked the pluripotent capacity of embryonic cells but were derived from adult tissue. "That's a very cool thing," Battey says. "Because now you have a source of pluripotent cells that doesn't require the problematic destruction of a human embryo."

But Obama was hardly likely to abandon embryonic-stem-cell research, for reasons of both science and politics. Although Thomson, and many others, had hoped that advances in adult stem-cell research would end the controversy by removing embryos from the equation, questions remained about the efficacy of the approach. When I asked Collins about the Thomson-Yamanaka breakthrough, he said that not enough was yet known about such cells to guess whether they have the same therapeutic potential as embryonic stem cells. For example, scientists have learned that the pluripotent cells derived from adult tissues retain some memory of that tissue. "Will that matter for the therapeutic uses we all dream of?" he asked. "No one knows, but it would be foolish now to proceed without comparing them at every step to the gold standard for pluripotency—and that remains the human embryonic stem cell. So it's not 'either/or' that we should be pursuing. It's 'both/and.'"

Politically, advocacy of embryonic-stem-cell research had become a reliable Democratic weapon against Republicans (the 2004 Democratic Vice-Presidential nominee, John Edwards, had vowed that if the Kerry-Edwards ticket won "people like Christopher Reeve will get up out of that wheelchair and walk again"), and was perhaps even decisive in key congressional elections in 2006. As a candidate in 2008, Obama himself had promised to overturn the Bush restrictions, part of his vow to return science to its proper standing.

Before Collins had a direct say in the Administration's decision on stem cells, he was personally torn by the ethical questions posed by stem-cell research. He has long opposed the creation of embryos for the purpose of research. He sees a human embryo as a potential life, though he thinks that it is not possible scientifically to settle precisely when life begins. But Collins also feels it is morally wasteful not to take advantage of the hundreds of thousands of embryos created for in-vitro fertilization that ultimately are disposed of anyway. These embryos are doomed, but they can help aid disease research.

The policy that Obama settled on, and Collins endorsed, seemed an elegant solution to the stem-cell controversy: the ban on funding new embryonic-stem-cell lines would be lifted, while keeping the Bush restrictions guiding how those new lines would be derived. Adult-stem-cell research projects, meanwhile, would be aggressively funded, at a rate more than twice that of the politically and ethically problematic research involving embryos.

This solution was widely applauded. Left unaddressed, however, was the potential snare of the Dickey-Wicker rider, which Congress, despite Obama's new policy, renewed again in 2009.

COLLINS PUBLISHED ANOTHER BOOK THIS year, *The Language of Life*, a companion to *The Language of God*, but this time the subject was genomics, and the great promise the field holds for biomedicine. Collins has long been the most enthusiastic, and persuasive, public champion of genomics, from the time when he was selling the Human Genome Project to Congress and defending the project against private-sector competition. ("There is no other scientific enterprise that humankind has mounted in an organized way that compares to this," he said then. "I am sure that history will look back on this in a hundred years and say, 'This was the most significant thing humankind has tried to do scientifically.'") Once the genome was sequenced, Collins directed his enthusiasm toward the medical "revolution"—his term—that would result. The ability to see "misspellings" in an individual's DNA, Collins proposed, would allow science to identify an individual's disease risks well in advance of onset; and understanding disease at the most molecular level would allow biomedicine to design therapies suited to that individual, moving medicine forward from its one-size-fits-all approach to disease. "If you fall ill," he wrote, "the therapeutic options waiting for you, many derived from new understanding of the human genome, will be both more effective and less toxic than the treatments available just a few years ago." This new "personalized medicine," Collins says, is the big prize that has always motivated his work.

Collins was disappointed that although health care dominated the political conversation for more than a year, the personalized-medicine agenda was scarcely mentioned. If anything, the focus of the reform effort in Congress was away from personalized medicine, and toward such cost-saving efforts as comparative-effectiveness research, which tests competing therapies on broad groups of patients and endorses the treatment that is effective on the largest group. Collins publicly cautioned against this approach last fall, and was joined by advocacy groups representing victims of various diseases. His caution was unheeded, and the health-care-reform law that eventually passed created a comparative-effectiveness bureaucracy.

Collins's greater disappointment, though, was the pessimistic tone attending the observation of the ten-year anniversary of President Clinton's announcement that the human gene code had been cracked. In June, the *Times* published on its front page an article headlined "A Decade Later, Genetic Map Yields Few New Cures." Collins called it "a pretty gloomy and cynical story."

Craig Venter, who was Collins's private-sector competitor in the race to unlock the human genetic code, and stood alongside Collins and Clinton at that sunny White House celebration ten years ago, disagrees. He estimates the medical benefits derived from the human genome to be "close to zero." In an interview with *Der Spiegel*, Venter said that decoding the genome has so far yielded nothing more than an increased understanding of an individual's probabilities of contracting a disease. "How does a one or three per cent increased risk for something translate into the clinic?" he asked. "It is useless information." As for a new era of personalized medicine, Venter said, "That was another one of these silly naïve notions that was out there. It's not 'Oh, we know your genome, we're going to make this drug for you.' That will never happen. It is more important that you use the information in the genome about your personal risks and reduce them through intelligent behavior."

Collins strongly disputes that assessment. He says that after reading the *Times* story he sat down and wrote out a list of breakthroughs directly attributable to the advances in genomics, among them providing new understanding of age-related macular degeneration, Crohn's disease and the role of autophagy, and Parkinson's disease and the central role of alpha-synuclein aggregates; and the development of a recent drug for lupus. "It's revolutionized everything that we do," he says. He has discussed some of this with his friend the militant atheist Christopher Hitchens: "As you might have heard, Christopher has esophageal cancer, and I have actually been spending a fair amount of time with him and his wife, Carol, trying to help him sort through the options for therapy—including some rather cutting-edge approaches based on cancer genomics."

Collins concedes that the prospects of sudden, practical benefits from the genome were initially overstated by some, and may still be a decade or more away. "You know about the first law of technology," he says. "A technological advance of a major sort almost always is overestimated in the short run for its consequences—and underestimated in the long run."

To Collins, and to most researchers, the controversy over embryonic stem cells seemed a settled matter. Last December, he announced that the first new cell lines had been approved, and by last week the total number of embryonic-stem-cell lines approved for federal research had reached seventy-five. This summer, the Food and Drug Administration approved the first authorized test in humans of an embryonic-stem-cell therapy. The cells, developed by the Geron Corporation and the University of California at Irvine, would be injected into the injury sites of spinal-cord patients, in the hope of restoring sensation and possibly allowing movement of paralyzed limbs.

Last week, Collins said that embryonic-stem-cell research had become "one of the most exciting areas of the broad array of engines of discovery that N.I.H. supports." Of Judge Lamberth's injunction against the Obama policy of funding the research, he commented, "This decision has just poured sand into that engine of discovery."

The effect of the ruling was dramatic, and almost immediate. Collins and his staff spent much of the week notifying researchers in labs across the country that, in some cases, their work would have to shut down within days. Researchers who had already received funding could continue their work, but would not be able to apply for renewed funding. Fifty promising projects that were up for peer review were pulled. Another dozen projects, which had scored highly in peer reviews and were awaiting final approval, were suspended. Twenty-two grants that were coming up for annual renewal in September were frozen.

Collins said that he was stunned by Judge Lamberth's decision, as were most people in the research community. Perhaps they should not have been. The ruling didn't arrive out of the blue. Although the battle over embryonic stem cells had receded from public contention, the issue had not gone away. Earlier this summer, a lawsuit against the Administration filed by a group of plaintiffs (including two researchers specializing in adult stem research), claiming that the new Obama rules harmed their work by increasing the competition for federal research dollars, was reinstated by a federal appeals court. That sent the case back to Judge Lamberth's court, where the plaintiffs asked for an injunction stopping the funding of embryonic-stem-cell projects. Lamberth granted the injunction, based not only on the question of whether the plaintiffs have been harmed but on his finding that the Obama policy, on its face, violates the Dickey-Wicker amendment.

Dickey-Wicker bans the use of public money for "(1) the creation of a human embryo or embryos for research purposes; or (2) research in which a human embryo or embryos are destroyed, discarded, or knowingly subjected to risk of injury or death." Lamberth didn't accept the Clinton argument, essentially adopted by Obama, that the embryo-killing stem-cell-extraction process could be separated from the subsequent research using the stem cells. "Congress has spoken to the precise question at issue—whether federal funds may be used for research in which an embryo is destroyed," Lamberth wrote. "As demonstrated by the plain language of the statute, the unambiguous intent of Congress is to prohibit the expenditure of federal funds on 'research in which a human embryo or embryos are destroyed.'"

The Obama Administration immediately decided to contest Lamberth's ruling, but, even if a stay of the injunction is secured, the research it concerns will remain in doubt until the underlying complaint is resolved, a determination in which the plaintiffs, according to Judge Lamberth, "have demonstrated a strong likelihood of success."

Blame for this new crisis in stem-cell research can, in large measure, be placed on the attenuated way that Congress, and even the White House, has dealt with the inherent politics of the issue. When Obama announced his new policy, one of those standing beside him was Democratic Representative Diana DeGette, of Colorado, a long-time champion of stem-cell research. During the Bush Administration, she co-authored (with the Delaware Republican Mike Castle) legislation that would have expanded embryonic-stem-cell research to roughly the extent that Obama later adopted. The legislation passed twice, and was twice vetoed by Bush. When Obama created his new policy, through an executive order, DeGette was worried that the Dickey-Wicker amendment, unaddressed, would someday rear its head.

"A lot of us knew that the Dickey-Wicker amendment was out there—we knew that it was an issue," DeGette told me last week. "But we also knew that, politically, it would probably be difficult for us to repeal Dickey-Wicker, because the pro-life members of Congress felt it served some purpose."

DeGette said that after the announcement of Obama's new policy, which she welcomed, she wanted to press ahead one more time with her bill, which would have effectively made the Obama rules law. It seemed the perfect opportunity—Republicans were in retreat, Obama's approval ratings were at their highest, and the Democrats, who controlled both branches of Congress, were feeling expansive. "I was all ready to go at that moment with the legislation," she says. "And what happened was—you know, this happens in political institutions—we had so many other things going on. If you'll remember, in March, 2009, we had the stimulus bill, we had the food-safety bill, we had the climate-change bill, and the health-care legislation. And so it kind of fell low on the agenda, particularly in the Senate."

The political moment is much less propitious for Democrats now, of course, but DeGette says that the shock of Judge Lamberth's ruling may alter the political arithmetic. She has reintroduced her bill, again co-sponsored by Castle, and may adjust it to address the Dickey-Wicker problems that Lamberth cited. She spent much of the summer pressuring the new Democrats in Congress who have never voted on the issue, and she thinks there is reason for hope. "Most of the Democratic freshmen, who are in vulnerable seats, are supportive of this legislation," she said. "In fact, a lot of them are running against anti-stem-cell Republicans, so they actually think it might help them in their reëlection, just as it helped a lot of Democrats in 2006." DeGette also said that her Republican allies are still with her, including Mark Kirk, the Republican congressman who is running for Barack Obama's Senate seat.

Whether or not DeGette's optimism is warranted, Francis Collins's bureaucratic skills will be severely tested in the coming struggle. Last week, his resolve was evident. "This goes beyond politics," he declared. As he put it to me, "Patients and their families are counting on us to do everything in our power, ethically and responsibly, to learn how to transform these cells into entirely new therapies. It's time to accelerate human-embryonic-stem-cell research, not throw on the brakes."

ANDREW CURRY

The Mathematics of Terror

FROM *DISCOVER*

Strange as it may seem, in many ways humans in groups behave like quantum particles. As Andrew Curry learns, this similarity applies to warfare, too.

SINCE THE 1960S, THE MOUNTAINS OF SOUTHERN Colombia have been home to a war between the government and a leftist guerrilla movement known as the Revolutionary Armed Forces of Colombia, or FARC. The conflict has simmered for decades. Sometimes it flares up in battles with government forces, a terror bombing, or a particularly high-profile kidnapping. Sometimes it fades into the background as cease-fires or negotiations quiet the hostilities. FARC has been fighting for so long that the war has become almost like background noise, says Neil Johnson, a University of Miami physicist who travels to Colombia every year to visit his wife's family. Even locals have become numb to the conflict. "There's this war going on, but I didn't think too much of it. You hear numbers of dead every day, like football results," Johnson says. "It took me 10 years to realize that maybe there was important information hidden in those numbers."

Johnson, who specializes in the study of complexity, is one of a new breed of physicists turning their analytical acumen away from subatomic particles and toward a bewildering array of more immediate human problems, from traffic management to urban planning. It turns out that subatomic particles and people are not that different, he explains. "The properties of individual electrons have been known for many years, but when they get together as a group they do bizarre things"—much like stock traders, who have more in common with quarks and gluons than you might think. So profound is the connection that quants (quantitative analysts, often with backgrounds in physics or engineering) have flocked to Wall Street, creating elaborate models based on the way markets have moved in the past. ArXiv, a clearinghouse for physics research papers, includes an entire section on "quantitative finance."

Still, it was not until a chance 2001 meeting in Bogotá with Mike Spagat, an economist at Royal Holloway College, University of London, that Johnson considered modeling something as human as warfare. Spagat had a Colombian Ph.D. student named Jorge Restrepo who was gathering data on attacks and death tolls, provided by the nonprofit Center for Investigation and Popular Education, so he could look for patterns in the conflict. Johnson hoped the numbers could tell them something about how the individual particles— in this case, insurgents rather than electrons—functioned when put together in large groups.

Soon the new team had a database that included more than 20,000 separate incidents from two and a half decades of FARC attacks. Johnson and Spagat expected that the success of the attacks, measured in the number of people killed, would cluster around a certain figure: There would be a few small attacks and a few large ones as outliers on either end, but most attacks would pile up in the middle. Visually, that distribution forms a bell curve, a shape that represents everything from height (some very short people, some very tall, most American men about 5'10") to rolls of the dice (the occasional 2 or 12, but a lot of 5s, 6s, and 7s). Bell curves are called normal distribution curves because this is how we expect the world to work much of the time. But the Colombia graph looked completely different. When the researchers plotted the number of attacks along the y, or vertical, axis and people killed along the x, or horizontal, axis, the result was a line that plunged down and then levelled off. At the top were lots of tiny attacks; at the bottom were a handful of huge ones.

That pattern, known as a power law curve, is an extremely common one in math. It describes a progression in which the value of a variable (in this case, the number of casualties) is always ramped up or down by the same exponent, or power, as in: two to the power of two (2 x 2) equals four, three to the power of two (3 x 3) equals nine, four to the power of two (4 x 4) equals 16, and so on. If the height of Americans were distributed according to a power law curve rather than a bell curve, there would be 180 million people 7 inches tall, 60,000 people towering at 8'11", and a solitary giant as tall as the Empire State Building. Although power laws clearly do not apply to human height, they show up often in everyday situations, from income distribution (billions of people living on a few dollars a day, a handful of multibillionaires) to the weather (lots of small storms, just a few hurricane Katrinas).

In Colombia's case, decades of news reports confirmed that the number of attacks formed a line that sloped down from left to right. In general, an attack that causes 10 deaths is 316 times as likely as one that kills 100. The larger the event, the rarer it is.

At first the pattern seemed too clear and simple to be true. "Immediately I thought, 'We need to look at another war,'" Johnson says. With the U.S. invasion of Iraq in full swing, he and his collaborators had an obvious second test. In 2005, using data gleaned from sources like the Iraq Body Count project and iCasualties, a Web site that tracks U.S. military deaths, they crunched the numbers on the size and frequency of attacks by Iraqi insurgents. Not only did the data fit a power curve, but the shape of that curve was nearly identical to the one describing the Colombian conflict.

Around that time, a Santa Fe Institute computer scientist named Aaron Clauset was applying the same approach to what seemed like a distinctly different problem. Rather than looking at specific guerrilla movements, Clauset was examining total deaths caused by global terrorist attacks since 1968. When he plotted nearly 30,000 incidents on a graph, they formed a curve to the power of -2.38. (The power number is negative because it reflects a decrease rather than an increase in the number of events as death tolls rise.) With its characteristic downward slope, the curve was eerily similar to those generated by Johnson and Spagat for Colombia and Iraq.

To rule out coincidence, Johnson, Spagat, and University of Oxford physicist Sean Gourley gathered data on nine other insurgencies. One after another, the curves clicked into place: Peru's Shining Path guerrilla movement: a curve with a power of -2.4. The Indonesian campaign against rebels in East Timor from 1996 to 2001: -2.5. The Palestinian second intifada: -2.55. Fighting against Afghanistan's Taliban from 2001 to 2005: -2.44. By contrast, traditional conflicts in which two armies squared off against each other (such as the Spanish and American civil wars) yielded graphs that looked a lot more like bell curves than power curves. Although the politics, religion, funding, motives, and strategies of the insurgencies varied, the power trends did not.

In an age of biological weapons and dirty nukes, the implications are chilling. Although truly massive power-law events—like the Great Depression or killer storms—are drastically less common than smaller disruptions, they still occur. In the normal distribution of a bell curve, you never get such extremes, but the pattern underlying the power curve enables a few rare events of extraordinary magnitude. One might use the math to argue that the 9/11 attack that killed more than 2,700 people in New York City was bound to happen. And there is ample reason to believe that an even bigger one is on the way, sooner or later.

For Johnson, a Cambridge- and Harvard-educated physicist who has studied stock markets and other apparently unpredictable systems, the power law was familiar territory. Whether in New York, Tokyo, or London, markets tend to follow the same boom-bust cycles, with little daily upticks and downticks punctuated every few decades by a big crash or boom. "Markets move every day, but some days they move a lot," Johnson says. "There are different people, different stocks, but that just seems to be the way people get together and trade."

If physics-based models can predict the behavior of stock markets, Johnson reasoned, why couldn't they foresee the behavior of insurgents so that attacks could be prevented? "Prediction is the holy grail everyone is in pursuit of," says Brian Tivnan, a modeling expert at a U.S. Department of Defense–funded think tank called the Mitre Corp. Tivnan brought Johnson's work to the attention of Pentagon officials. "We were very encouraged to see physicists and mathematicians looking at the data from an apolitical, analytic perspective," he says.

But if they were going to develop a predictive model, Johnson and his team would have to figure out what it was about the behavior of insurgents and terrorists that made their bloody fingerprints so similar all around the world. They started by tossing the traditional take on insurgencies out the window.

Conventional counterinsurgency thinking tries to get into the heads of rebels by understanding their motivations and methods.

Political scientists and sociologists studying the conflicts in Iraq and Afghanistan have emphasized tribal affiliations, nationalism, religion, social networks, and other cultural concerns. Using lessons learned (or perhaps mislearned) in Vietnam, meanwhile, Pentagon planners approached these conflicts as if they were facing smaller armies with worse equipment, hoping that if they could knock out the enemy's leadership they would decapitate and demoralize the insurgency.

But these assumptions were off. Guerrilla fighters in Vietnam, like U.S. troops, answered to a central command; insurgents in Iraq did not. And from a physics point of view, getting inside an insurgent's head was irrelevant. "In political science literature, human rationality is primary. They assume groups are rational actors, have access to all the information, and make the right decisions," Clauset says. "A physicist's natural approach is to assume people are like particles, and their behavior the result of constraints beyond their control."

Basing their computer models on programs written to predict all sorts of fluctuating phenomena, from traffic flow to stock prices, Johnson's team tried to create equations that reflected the behavior of the individual insurgents seen in the data. The equations that came closest "involved a soup of conflict groups of varying strengths, in a constant process of coalescing and dissolving," Spagat says.

Johnson likens the insurgent groups in his computer model to a pane of glass that shatters into smaller and smaller splinters with each hit. The bigger shards are capable of delivering the deepest, nastiest cuts, but they are also the easiest to target. The smallest slivers of glass, on the other hand, might deliver the casualty equivalent of a pinprick, but there are so many of them, and they are so hard to spot, that the total amount of damage they cause stays high.

If the model is correct, then insurgents conduct "asymmetrical warfare," battling a larger and better-equipped enemy with a loose network of fighters lacking central command. However obvious this seems today, it was a concept that escaped American military planners when the fighting in Iraq and Afghanistan began nearly a decade ago. "The insurgents kept the most powerful military the world has ever seen at bay for four years," says John Robb, a former Special Operations pilot and author of *Brave New War: The Next Stage of Terrorism and the End of Globalization.* "You're not going to defeat them by killing groups or killing people. You have to change the entire dynamic. It's a tough lesson for a lot of military folks to absorb." Indeed, the harder the U.S. forces hit, the more the insurgency shattered into near-invisible shards. By the time Johnson's paper was published in *Nature* last year, the military had learned, through bitter experience, the futility of fighting insurgents with traditional tactics. (The military has never published on the issue, but Johnson says that strategists have recently heard about his ideas.)

The splintered, disorganized nature of insurgencies became still clearer when Johnson and his colleagues looked at the timing of attacks. The numbers in Iraq, Colombia, Peru, and Afghanistan followed similar patterns, with "sudden bursts of activity, then quiet periods," Spagat says. "If it were random, you would have far fewer busy days and far fewer quiet days than are captured in the data." Without a centralized command to issue orders, there must be something else behind the clustered timing of attacks.

Spagat and Johnson argue that the missing element is the role played by media and other sources of information. For an insurgent group, a successful strike is not one that does the most damage, but one that draws the most attention. "Media and publicity are the oxygen of terrorism," says anthropologist Scott Atran, an expert on terrorism at the National Center for Scientific Research in Paris. "Without them, it would die."

Spagat likens the relationship between the dozens of groups in Iraq and the media to drivers at rush hour. Much as drivers try to outguess other drivers to pick the least-traveled route home, the data suggest that terrorists and insurgents aim to stage their attacks when they will have the media's undivided attention. "Instead of competition for road space, there's competition for media space," he says. "You want to be the only person making an attack on a given day. If there are more than a few attacks on a given day, your story tends to get lost in the system."

But since there is no one to coordinate attacks, the resulting patterns are "bursty," a term used to describe many real-world events that unfold in short, intense fits. Think of the traffic jams that seem to come out of nowhere and disappear just as quickly. They are the product of thousands of drivers with incomplete information trying to outguess thousands of other drivers trying to pick the best route home. Sometimes enough people will guess wrong and spend two hours sitting on the freeway.

As FASCINATING AS THEIR MATHEMATICAL patterns are, Johnson and Spagat remain far from their goal: anticipating attacks and being able to stop them. Atran says the researchers' findings bear out what he has seen during his fieldwork on the psychology of suicide bombers and the importance of media attention. But that level of understanding is not good enough. In the end, the math may not explain it all, he contends. "Insurgencies are sui generis; each takes place within its own social, cultural, and political milieu. Trying to create a unified model is a fool's errand. I don't think there is enough cultural awareness of what moves people to do what they do."

Cultural context is not something Johnson pays much attention to. Accustomed to analyzing particles, which are not known for their reasoning capabilities or complex inner lives, physicists tend to ignore the why and go straight to the how. "All those questions of 'why' show a lack of understanding," Johnson insists. "Whatever the reasons are, this is how they operate." He has explained this to British and American military officers, Iraqi officials, and even security officials at the London Olympics. "Insurgents may be doing it for all sorts of reasons, but the mechanics are what matters."

That kind of talk makes many counterinsurgency analysts bristle. Andrew Exum, a fellow at the Center for a New American Security who led a platoon of Army Rangers in Iraq and Afghanistan, says that quantitative analysis is a useful tool, but only when it is sensitive to the complexities of real-life situations and is coupled with the expertise of someone well versed in the specific political and religious contexts at hand, as well as in military strategy. "I'm turned off by the confidence with which these scholars presented their model," he says. "A little more humility might have been in order."

The complaint has definite resonance in the wake of the recent financial crisis, which saw Wall Street quants creating ostensibly rational models that drove the financial markets to the brink of disaster. While academic physicists like Johnson try to account for the behavior of the traders in their models, the standard quant approach is based on markets moving at random. It is an approach that Johnson is eager to distance himself from. "If you account for human collective behavior, you get results that are different from the standard quant models," he says. "We started looking at financial models precisely because we thought crashes were not properly taken into account." From a distance the difference can seem academic; to most people, a computer model is a computer model. Johnson admits he has had as little luck selling his power law approach to firms on Wall Street as to traditionalists in the military.

Former Special Operations pilot Robb, an advocate of the mathematical approach, says the cool reaction to the quantitative analysis of terrorism is par for the course. During the Vietnam War, soldiers blamed the number crunchers—those informing decisions in the Pentagon based on body counts and kill ratios—for the war's bad turn. As a result, "a lot of people think counterinsurgency is very qualitative, very mushy, and should stay that way. It's almost a mystical thing," Robb says.

"Nothing we've done suggests we can predict there will be an attack in, say, the next two weeks," Spagat freely concedes. "Rather, a physics-inspired insurgency model can help guide more general decisions. If the data show that attacks happen in a bursty pattern, it makes sense to have emergency medical teams able to react to several attacks at once. And the data offer a rough guide to how big those attacks might be, based on how they've looked in the past." Moreover, he says, if the model is right about modern insurgencies' being a constantly shifting collection of small, unconnected groups, it would be a useful tool for military planners trying to find the most effective tactics. Notable military research groups such as Mitre and the Pentagon's IED Defeat Organization have met with Johnson and Spagat to talk about their work. During the course of such meetings, Johnson must counter the ingrained notion that human behavior is uniquely complex and unpredictable.

Never before have researchers had ready access to decades' worth of social data that could be analyzed, and never before has it been so easy to find patterns amid the complex streams of numbers. As the world learned after the Wall Street crash, finding patterns is not the same as understanding which ones are meaningful and acting on them in a responsible way. But given the rush of numbers, the analytical approach of physicists and economists—on Wall Street and now in war—will inevitably keep spreading, Clauset says, "We are entering an era in which social sciences have access to a wealth of data beyond their wildest dreams."

About the Contributors

Working in such diverse fields as government, politics, finance, fashion, sports, and entertainment, JUDY BALABAN (Quine) has served on nonprofit boards and as a consultant for several organizations addressing civil rights, civil liberties, the arts, education, and veterans issues. A mother of three with four grandchildren, she resides in Beverly Hills, California; has written articles for magazines, trade journals, and other books; cowritten *Superlife,* a mind-body-spirit exercise book with actress Stefanie Powers; and is author of a partially autobiographical nonfiction bestseller, *The Bridesmaids: Grace Kelly, Princess of Monaco, and Six Intimate Friends.* CARI BEAUCHAMP is the award-winning author of five nonfiction books, including *Without Lying Down* and *Joseph P. Kennedy Presents.* She is a contributing writer for *Vanity Fair,* has written documentary films, and has twice been named an Academy of Motion Pictures Arts and Sciences Film Scholar.

"While we wrote, and after our article appeared in *Vanity Fair,*" they report, "we were pleased to be in touch with scientists conducting research with LSD and similar compounds. Today, doctors in several countries are using LSD and creating offshoot products to assist people suffering from a broad variety of medical problems including end-of-life issues, PTSD, and cluster headaches. It was this potential—and not its use as a street drug—that drew the celebrity patients we wrote about into government-licensed LSD25 experimental psychotherapy so many decades ago."

BURKHARD BILGER writes about food, science, and American subcultures for *The New Yorker,* where he has been a staff writer since 2000. His work has also appeared in *The Atlantic Monthly, Harper's,* the *New York Times,* and numerous other publications and anthologies. Bilger was a senior editor at *Discover* from 1999 to 2005. Before that, he worked as a writer and deputy editor for *The Sciences,* where his work helped earn two National Magazine Awards and six nominations. His book, *Noodling for Flathead,* was a finalist for the PEN/Martha Albrand Award in 2000.

"I first heard of Sandor Katz, the main character in 'Nature's Spoils,' while visiting an underground bread club in Oregon," he recalls. "I'd always been fascinated by illegal foods and the back-and-forth between science and tradition in matters of hygiene. Sandy seemed to be at the heart of this discussion, and of a sprawling network of antiestablishment foodies. He was also a wonderful cook—always a boon when reporting a story. I wish the same could be said of everyone I met while I was with him. . . ."

DEBORAH BLUM is a Pulitzer Prize–winning science writer and the author of five books, most recently *The Poisoner's Handbook: Murder and the Birth of Forensic Medicine in Jazz Age New York.* She writes for publications from the *New York Times* to *Slate,* and her blog, Speakeasy Science, is part of the Public Library of Science network. She teaches writing at the University of Wisconsin, where she is a professor of journalism.

"When I titled this piece 'The Trouble with Scientists,' " she says, "I was thinking about the old Alfred Hitchcock horror-comedy *The Trouble with Harry,* in which a corpse keeps reappearing and causing endless trouble. These conflicts between science journalists and scientists also keep reappearing and they can distract from the issue of good science communication. It's frustrating because science literacy is such a challenge in this country. Solving that problem, sharing the stories of science, should be the real priority. Oh, and plus, I've always found my father's reaction to my career choice hilarious. As you can tell by the end of the piece, he's actually a very good guy."

MARK BOWDEN is a bestselling author and journalist, best known for his book *Black Hawk Down,* a finalist for the National Book Award, and the basis for the film of the same name. He has written six others, and his latest, *Worm,* is scheduled for publication in the fall of 2011. He is a contributing editor to *Vanity Fair,* and a national correspondent for *The Atlantic.*

"I have the *Wall Street Journal* to thank for this story," he explains. "It ran a short article on its front page in early 2009 about the Conficker worm, and it struck me that I hadn't the vaguest notion what it was about. Something to do with a major threat to the Internet. I've been at this long enough to believe that if I don't understand a page-one story in the newspaper, few others do either. So I tore it off the page, stuck the clip in a manila folder, and started trying to find out."

PETER J. BOYER became a special correspondent at *Newsweek/DailyBeast* in 2011, after spending eighteen years as a staff writer at *The New Yorker,* where he wrote on a wide range of subjects, including politics, the military, religion, and sports. Before joining *The New Yorker,* Boyer was a reporter for the *Los Angeles Times* and the *New York Times,* a contributing editor at *Vanity Fair,* and a television critic for National Public Radio's *Morning Edition.* As a correspondent on the documentary series *Frontline,* he won a George Foster Peabody Award, an Emmy, and consecutive Writers Guild Awards for his reporting. Boyer's *New Yorker* articles have been included in the anthologies *The Best American Political Writing, The Best American Science Writing, The Best American Spiritual Writing,* and *The Best American Crime Writing.* He is at work on a book about American evangelism.

"I've always been fascinated by the intersection of science and faith," he admits, "but I'd never imagined seeing a smashup there. Then I watched the reaction of a very vocal segment of the scientific community to the elevation of Dr. Francis Collins to the top chair at the National Institutes of Health. Collins is one of the world's most accomplished scientists (having, among other achievements, led the Humane Genome Project), but, to some, his faith in Jesus Christ disqualified him for the NIH job. His handling of the tricky stem-cell-research issue proved them wrong."

JOHN BRENKUS has built BASE Productions into one of Hollywood's leading television production companies by allowing audiences to experience and appreciate the science behind the most extreme events, conditions, and experiences in the world. As creator and host of the multi–Emmy Award–winning *Sport Science,* Brenkus has broken new ground in the world of sports and television. By putting the world's greatest athletes to the ultimate test in the *Sport Science* lab, Brenkus has unique and fascinating insight into just how far away we are from achieving the limits of human performance.

"From the time that you're just old enough to swing a bat," he notes, "the most appealing aspect to baseball is trying to hit the ball as far as you possibly can. And with each passing major league season, it seems like the pros keep hitting them farther and farther out of the park. But what's the farthest a baseball can possibly be hit? In this chapter of *The Perfection Point,* I optimized all the factors that make a baseball sail out of the park and determine once and for all what is the limit to the homerun."

KATY BUTLER was born in South Africa, raised in England, and educated at Wesleyan University. She is a journalist, memoirist, and cultural critic whose writing has appeared in *The New Yorker,* the *New York Times, Mother Jones,* and *Tricycle, The Buddhist Review.* A prior finalist for a National Magazine Award, she lives in northern California and regularly teaches memoir at Esalen Institute. She is currently writing her first nonfiction book, *Knocking on Heaven's Door: A Journey Through Old Age and New Medicine,* for Scribner, due out in 2013.

She writes: "Before I watched my father's broken life prolonged by a hastily implanted pacemaker, I was a journalist interested mainly in memoir, addiction, neuroscience, and mental health issues—and naive about the overall workings of the American medical system. Ever since I began to write, however, I've used my personal experiences as lenses through which to view broader cultural issues.

"So after my father died, I could not rest until I understood the larger forces that had shaped his prolonged and attenuated dying. It was a difficult piece to write: in one paragraph I would I painstakingly record my parents' suffering, and in the next, explain the perverse economic incentives that reward medical overtreatment near the end of life. Sometimes my heart and mind were uneasily harnessed and jostling in the traces.

"I began the story a month after my father's death, and finished it the same week my mother died. The story struck a nerve: it was the fourth most accessed article in *The New York Times Magazine* for 2010, won a first-place award from the Association of Health Care Journalists, and was named a Notable Narrative by the Nieman Foundation at Harvard's project on narrative nonfiction. Perhaps more important, more than a thousand readers, including many troubled doctors, emailed me or the *Times* website with stories of their own, uncovering a well of uneasiness over our medicalized approach to the end of life, and inspiring me to write my first book."

JOHN COLAPINTO is a staff writer at *The New Yorker* and author of the *New York Times* bestselling nonfiction book *As Nature Made Him: The Boy Who Was Raised as a Girl* and the novel *About the Author.* His books have been translated into dozens of languages, both have been sold to the movies, and he has won the National Magazine Award for reporting. He is currently finishing a new novel, *The Ephebophiles.*

" 'Mother Courage' is a story I set out to report with real trepidation," he reveals, "given the tragic dimensions of the disease and the lack of a cure. But like Pat Furlong herself, I found such fascination in DMD, and scientists' efforts to understand and control it, that the story quickly became one of the most absorbing, and inspiring, of my career."

ANDREW CURRY (www.andrewcurry.com) is a foreign correspondent based in Berlin. A former Fulbright journalism fellow, he covers science, history, politics, and culture for magazines including *Discover, Smithsonian,* and *Wired,* and is an *Archaeology* contributing editor.

"As a writer, I love exploring the places where science and society collide," he says. "This story is a perfect example—an instance where physicists and economists try to make an unlikely contribution to the field of counterinsurgency using the quantitative, data-oriented method they know best.

"The results were predictable, given the reception similar approaches have had on Wall Street: The counterinsurgency 'practitioners' I talked to were sniffy and dismissive of the entire project. They found the idea that human behavior can be accurately modeled dangerously absurd.

"Johnson, Spagat, and Clauset didn't seem too bothered. Their main concern was expanding and revising their data set to refine and test their models. I think maybe they go to bed dreaming about what the Pentagon's classified incident reports might hold."

TIM FOLGER has been writing about science for more than twenty years. His work has appeared in *Discover, National Geographic, Scientific American,* and other magazines. In 2007 he won the American Institute of Physics science writing award.

"Some examples of climate change are dramatic and photogenic," he notes. "Melting glaciers, stranded polar bears. But quieter transformations are happening all around us, and those are the sorts of changes Dave Bertelsen has spent decades tracking in the Sonoran Desert: flowers blooming at higher elevations; the absence of birdsong on a winter morning. His work is impressive and disturbing."

CYNTHIA GORNEY, who teaches at U.C. Berkeley's Graduate School of Journalism, is a contributing writer for the *New York Times Magazine* and *National Geographic*. A former reporter for the *Washington Post,* she is the author of *Articles of Faith: A Frontline History of the Abortion Wars,* and has written for many national magazines, including *The New Yorker, Harper's, Sports Illustrated, Runner's World,* and *Mother Jones.* Her profile of the tennis player Rafael Nadal was included in *Best American Sports Writing 2010.*

She writes: "Personal research—my own perplexed mission to try to understand what was happening to my brain, with and without estrogen replacement—led me to the story at the center of this piece for *The New York Times Magazine.* It wasn't until after the story appeared, and I began hearing from women all over the country, that I came to appreciate how many other women had misunderstood the controversial Women's Health Initiative as thoroughly as I had, and had found themselves suffering greatly as a result."

AMY HARMON is a *New York Times* national correspondent who covers the impact of science and technology on American life. She has won two Pulitzer Prizes, one in 2008 for her series "The DNA Age," the other as part of a team in 2001. Her chronicle of a clinical trial, "Target Cancer," received several honors this year. Harmon's first journalism job was at the *Los Angeles Times,* where she parlayed her amazing early-1990s ability to send email into a beat. Her article "Facing Life with a Lethal Gene" appeared in *Best of the Best American Science Writing* in 2010.

"Truth be told, I did not visit the nursing home where I first met the robotic seal called Paro with an open mind," she confesses. "Just the idea of using the furry thing to soothe residents who might think it alive seemed creepy and immoral. When it cooed in my arms, I hardened my heart. As the faces around me lit up, I refused to melt. It took a veteran in a wheelchair to set me straight. Of course they knew it wasn't real, he replied to my oh-so-delicate questioning. If Paro was the first step on a slippery slope toward replacing human relationships with machine simulations, the implication was, so be it. Meanwhile, could I please lighten up? Chastened, I stroked the seal's matted white fur, and allowed myself to smile."

CHARLES HOMANS is features editor at *Foreign Policy* and a contributing editor at *The Washington Monthly.* His work has also appeared in *The New Republic, Slate, The Economist,* and *Bookforum,* and on several National Public Radio programs. He was a finalist for the 2010 Livingston Award for Young Journalists for his reporting on the future of NASA's human spaceflight program.

"The question of how scientific information moves through society is as interesting to me as the information itself, especially in climate science," Homans says. "'Hot Air' was an attempt to slice off a little piece of that question and try to answer it—and also to poke around in the subculture of TV weathercasters, which turns out to be hugely interesting in its own right."

KRISTIN OHLSON is the author of a memoir, *Stalking the Divine,* and coauthor of *Kabul Beauty School.* Based in Cleveland, she's written about a variety of topics for a variety of publications, including the *New York Times, Salon,* Smithsonian.com, *Gourmet, Utne Reader, Preservation, Discover, New Scientist,* and many others. She is a recent recipient of a Community Partnership for Arts and Culture fiction fellowship.

"When I was driving around southern Ohio a few years ago," she remembers, "someone told me you could see smoke from a coal mine fire rising in nearby Wayne National Forest. In 1884, a group of pissed-off striking coal miners sabotaged their worksite. They loaded coal cars with logs, set them ablaze, and pushed them into the mine. It caught fire and an estimated two hundred square miles of coal were burned; it still burns. I was completely fascinated by this story and, luckily, found geologist Glenn Stracher, who knows just about everything there is to know about coal fires."

MICHAEL S. ROSENWALD is a staff writer at the *Washington Post*. He is also a magazine writer whose work has been published in *The New Yorker, Esquire, Smithsonian, GQ, Bloomberg BusinessWeek, Popular Science, Garden & Gun, Tin House, Creative Nonfiction, Men's Journal,* and *ESPN the Magazine.* A former finalist for the National Magazine Award in feature writing, Rosenwald has edited an anthology of Gay Talese's sportswriting, *The Silent Season of a Hero.* He has an MFA in creative writing from the University of Pittsburgh.

"Since my investigation into my status as a potential hoarder—and the subsequent dismantling of my piles—I've gone back to my old ways," he admits. "My side of the bedroom is a minor disaster. My car is able to transport me but not any other humans. I plan to tackle the problem more seriously again soon. Until then, I am ever more thankful of my wife's continued tolerance."

ALAN SCHWARZ is a reporter for the *New York Times* who, from 2007 until 2011, reported primarily on the risks of head injuries in sports, particularly football. His work forced sweeping changes in the handling of concussions by the National Football League and was credited with making youth sports safer for all athletes. He was honored with a 2010 Polk Award and was a finalist for the 2011 Pulitzer Prize for Public Service.

"Looking back, months after writing this story," he reflects, "it's hard for me not to sense the empathy I must have had for Fred Mueller, in that having one's work revolve around dead and injured children can be dispiriting—but ultimately rewarding. For me, spending four and a half years pulling the curtain back on football head injuries—the comatose kids, the clueless parents, and the lying doctors—was at best unpleasant, at worst downright maddening. You do it so that no one else ever has to."

CHARLES SIEBERT is the author of three critically acclaimed memoirs, *The Wauchula Woods Accord: Toward a New Understanding of Animals, A Man After His Own Heart,* and *Wickerby: An Urban Pastoral,* a *New York Times* Notable Book of 1998, as well as a novel, *Angus,* and *The Secret World of Whales,* a children's book about whales. A poet, journalist, essayist, and contributing writer for the *New York Times Magazine,* he has written for *The New Yorker, Harper's Magazine, Vanity Fair, Esquire, Outside, Men's Journal,* and *National Geographic.*

"This story began as a fairly straightforward look at dog fighting," he explains. "But in the course of the research I found myself becoming increasingly intrigued by the consistent connection between activities like dog fighting and other violent crimes, in particular spousal and child abuse. This, invariably, led to an exploration of empathy and its absence or impairment in the minds of those who abuse animals—human or non- —and those who have themselves been abused. We've long known to be wary of animal abusers. Modern laboratory and social sciences, however, are now giving us unprecedented looks inside both the abusers' houses and heads."

MICHAEL SPECTER, a staff writer at *The New Yorker* since 1998, writes about science and global public health. His book, *Denialism,* published by Penguin, is out in paperback.

"I wanted to write about TB in India because it remains so deadly," he says, "but originally I thought I would focus on the interaction between private enterprise and public health. I was unaware of the astonishingly successful new diagnostics that have the potential to cure and prevent tuberculosis. If used widely, they could become as valuable as antibiotics and vaccines are in places like India. Happily, soon after the story was published, the World Health Organization endorsed the methods and now recommends these advanced diagnostics as the basic TB test in most of the developing world."

JULIA WHITTY is an award-winning author of fiction and nonfiction. Her books *Deep Blue Home, The Fragile Edge,* and *A Tortoise for the Queen of Tonga* were awarded the O. Henry Prize, John Burroughs

Medal, PEN USA, and Kiriyama Prize, and were nominated for the PEN Hemingway, Dayton Literary Peace Prize, Orion Book Prize, and others. Whitty is science and environmental correspondent at *Mother Jones* and a blogger at The Blue Marble and Deep Blue Home. A former filmmaker, her more than seventy nature documentaries aired on PBS's *Nature*, The Discovery Channel, National Geographic, Outdoor Life Channel, Arts & Entertainment, and other broadcasters worldwide.

"Early on," she writes, "I knew an important facet of BP's disaster would be the long-term revelations destined to emerge as scientific focus shifted to the Gulf of Mexico—a place overdue for deep investigation and deep understanding. Though media attention has drifted away from the story, I remain hopeful that the countless lines of examination launched last year will help us understand the Gulf's treasures better. Maybe even help us recalibrate our compass towards a cleaner energy future."

ED YONG is a British science writer. His blog Not Exactly Rocket Science won the National Academies Science Communication Award in 2010, and his other writing has appeared in *New* Scientist, the *Times* (London), *Wired*, the *Guardian* (U.K.), *Nature*, the *Daily Telegraph*, *The Economist*, and more. His other awards include the Research Blogging Award in 2010, the Association of British Science Writers Best Newcomer Award in 2009, and the Daily Telegraph Science Writer Award in 2007. He is an active part of the scientific community on Twitter (@edyong209) and gives regular talks on science journalism and blogging.

"To the nearest approximation, we are made of bacteria," he explains. "I find this idea to be deeply and perversely fascinating and, apparently, so does everyone else. Since writing this piece, scientists showed that these intestinal hitchhikers are different in African villagers versus European urbanites, in infants versus adults, and in Caesarean-section babies versus those born naturally. These microbes recap our evolutionary history, they affect our health, and they could even shape the development of our minds. And despite all of that, this story—the tale of genes that stumbled from ocean to bowel, and the tale of the scientists who stumbled across *them*—is still my favorite. It also won the Three Quarks Daily Science Prize in 2010, judged by Richard Dawkins."

CARL ZIMMER is the author of ten books about science, including *Parasite Rex* and *The Tangled Bank: An Introduction to Evolution.* "The Singularity" is included in his ebook, *Brain Cuttings: Fifteen Journeys Through the Mind* (Scott & Nix). He is a regular contributor to the *New York Times,* and writes regularly for magazines such as *National Geographic, Scientific American,* and *Discover,* where he is a contributing editor and writes a blog about nature called The Loom.

"My stories usually come out of ideas I bring to my editors," he says, "but this one emerged in the reverse direction. My editor had heard all sorts of incredible claims about the Singularity and wanted to find out more about it. I warned him I might turn in a critical article, and, fortunately, he was perfectly happy with that possibility. What I ended up with was not a hatchet job, however, but a reflection on the progress of science, and what we crave to get from the future."

Permissions

A Note from the Series Editor

Submissions for next year's volume can be sent to:

Jesse Cohen, Editor
The Best American Science Writing 2012
c/o HarperCollins Publishers
10 E. 53rd St.
New York, NY 10022

Please include a brief cover letter; manuscripts will not be returned. Submissions can be made electronically and sent to jesseicohen@ netscape.net.